PRAISE FOR
THE BODY POLITIC

"Moreno shrewdly tracks the history of science in American politics from Thomas Jefferson to today's science culture wars. He explains how science and discovery have been central to our vision for the country, but often fueled a significant counter reaction. A must read for anyone who wants to understand science policy today."

—JOHN PODESTA, President and CEO of the Center for American Progress

"*The Body Politic* reminds us that science occurs within a complex context that exerts powerful forces upon scientists, public officials, advocacy groups, and patients. Moreno has written the kind of book that needed to be written, combining detailed research, enlightened analysis, and an important message, all wrapped in accessible text."

—ERIC M. MESLIN, Ph.D., Director, Indiana University Center for Bioethics

"Moreno clarifies major points of science-society tension over the last half century and brings a sharp eye to the societal context confronting future advances and their applications."

—ALAN I. LESHNER, Ph.D., Executive Publisher, *Science*

"A beautiful book."

—JAY SCHULKIN, Research Professor, Department of Neuroscience, Georgetown University

"*The Body Politic* is required reading for anyone who wants to understand the history of American political thought about science, the dynamics of current controversies such as the stem cell debate, and the battle between those who see science as the route to a better future and those who see within science the potential for a loss of our sense of human distinctiveness and dignity."

—PAUL ROOT WOLPE, Ph.D., Director, Center for Ethics, Emory University

"This groundbreaking must-read book situates the biological revolution in its historical, philosophical and cultural context and, with almost breathtaking elegance, shows how society may come to define itself by the body politic."

—NITA A. FARAHANY, Associate Professor of Law & Associate Professor of Philosophy, Vanderbilt University

ALSO BY JONATHAN D. MORENO

Progress in Bioethics: Science, Policy, and Politics

Science Next: Innovation for the Common Good from the Center for American Progress

Mind Wars: Brain Research and National Defense

Is There an Ethicist in the House? On the Cutting Edge of Bioethics

Undue Risk: Secret State Experiments on Humans

Deciding Together: Bioethics and Moral Concensus

THE BODY POLITIC

The Battle Over Science in America

JONATHAN D. MORENO

BELLEVUE LITERARY PRESS
NEW YORK

First published in the United States in 2011 by
Bellevue Literary Press, New York

FOR INFORMATION ADDRESS:
Bellevue Literary Press
NYU School of Medicine
550 First Avenue
OBV 612
New York, NY 10016

Bellevue Literary Press would like to thank all its generous donors, individuals and foundations, for their support.

Library of Congress Cataloging-in-Publication Data

Book design and type formatting by Bernard Schleifer
Manufactured in the United States of America
FIRST EDITION
10 9 8 7 6 5 4 3 2 1
ISBN 978-1-934137-38-3 pb

In memory of

Marcia Lind

and

Steve Sussman

To change one's life:
Start immediately.
Do it flamboyantly.
No exceptions.

—William James

CONTENTS

PREFACE

I STARTED TEACHING AND WRITING ABOUT BIOETHICS MORE THAN THIRTY years ago, just after the birth of the first "test tube" baby. I finished this book just as the embryonic stem cell debate was reignited. What were once exotic scientific procedures have become familiar, but public concern about where modern biology and medicine are taking us is greater than ever. Though the underlying ethical issues about the ways we learn about and manage life and death cannot be avoided, they need to be understood in a broader social and historical context. My goal in this book is not to suggest final answers to the ethical questions, but to develop a larger historical, philosophical, and cultural framework that can enrich the moral conversation.

Some readers may be surprised that a professor of medical ethics is interested in "biopolitics." Isn't politics incompatible with the serious study of ethics? The fact that the word politics is often presumed to be pejorative is a sad commentary on the temper of our times. In the final analysis, politics is the way human beings organize their lives together. Considering all the challenges life presents, this is not a simple task and when one steps back what's amazing is that, on the whole, we're able to work things out. A serious look at where we've succeeded and failed, both politically and morally, can't help but be useful for understanding both politics and ethics.

History has to start somewhere. In that sense it is radically located, as is each of us, in a particular time and place. Biopolitical issues are necessarily themselves contextual: they have a particular flavor and a certain historical dynamic. How does science, and especially modern biology with all its attendant controversies, fit into that dynamic? My experiences in the policy world and in academia have led me to conclude that a new biopolitics is emerging and has been doing so for some time, though it has not been adequately recognized or assessed. Recognition of that fact is healthy and can help us understand some of the forces swirling around us. Besides the power of the new biology, both real and symbolic, the reasons for the appearance of the new biopolitics are many and complex. Assuredly, I bring a certain point of view to the issues themselves, but at the end of the day what I find most compelling are the cultural factors that have led us to this point.

Americans especially have a particular relationship to the ideas of science and progress. The ways we respond to the implications of modern biology are of great importance for the country at many levels: for the future of our economy, our place in international technological innovation, our sense of national purpose, the social and ethical choices that await us, and our self-understanding as a people. As the child of immigrant parents, I'm acutely sensitive to the fact that Americans are tied together by a common future rather than a shared past. For this reason, Americans are especially eager for an answer to the question, "Who are we?" Because it will help determine what idea of the future we measure ourselves against, the new biopolitics now unfolding will be part of an answer to that question for the twenty-first century.

ACKNOWLEDGMENTS

William James said that to every new experience we bring to bear all of those that have gone before. While writing this book I have often felt that way. My debts begin with those who first taught me American intellectual history more years ago than I care to admit. More recently I have learned from and been inspired by my wonderful colleagues at the University of Pennsylvania in Medical Ethics and History and Sociology of Science. John Tresch's work on Poe and science led me to one of the epigrams. I am grateful to Penn's president, Amy Gutmann, for her personal support and friendship and for conceiving the "Penn Integrates Knowledge" faculty group of which I am fortunate to be a part.

Many of the ideas in this book were developed through presentations to dozens of colleges, universities, professional organizations, and civic groups over the past eight years. For their hospitality and patience I am grateful to all of them. In the fall of 2009 I offered a course on biopolitics at Penn that helped me sharpen much of the discussion that appears here.

My introduction to the world of public policy and the opportunity to participate in a very modest way in shaping it are owed to John Podesta and colleagues at the Center for American Progress. To work with the dedicated people at organizations like CAP is to be immunized against cynicism about the policy process; I wish all Americans could have such opportunities. I have learned much from three young American Progress staff assistants: Sam Berger, Andrew Pratt, and Mike

Rugnetta. With such smart, hard-working, and dedicated people like them in the next generation of leaders, our country will fare well.

Smart and civil disagreement is the mother's milk of improved thinking about complex issues. Some of those whose views are different from mine are cited in this book, others have had less direct but no less important effects on my thinking. I am grateful to them for trying to set me straight and assure them that I am still listening.

Particular thanks are owed to Andrew Hogan of Penn, Paul Lombardo of Georgia State University and John Arras of the University of Virginia for helping me to avoid some of the errors in earlier drafts of this book. Ilana Yurkiewicz both helped me organize the endnotes and applied a keen and fresh critical eye to the text. Any remaining errors are of course my responsibility alone.

What a pleasure it is for a writer to have an editor like Erika Goldman. Erika is not only the best in the business but has been my friend, confidante, and guide to the mysterious publishing world for a dozen years. Even beyond her sensitivity and insight, she has a gentle way of guiding writers (especially this one) back on track.

As ever, I'm singularly grateful to Leslye, Jarrett, and Jillian for giving my life a center and for tolerating the odd schedule and emotional ups and downs of an author trying to discern the shape of the beast.

Portions of chapter 4 appeared in Jonathan D. Moreno and Sam Berger, "Biotechnology and the New Right: Neoconservatism's Red Menace," *American Journal of Bioethics* 7, no. 10 (2007): 7–13.

Natalie Ram helped prepare some of the material on chimeras and hybrids in chapter 6.

Portions of chapter 7 appeared in Arthur L. Caplan and Jonathan D. Moreno, "The Havasu 'Baaja Tribe and Informed Consent," *The Lancet*, 376, no. 9736 (2010): 141–204.

—JONATHAN D. MORENO

The rapid Progress *true* Science now makes, occasions my regretting sometimes that I was born so soon. It is impossible to imagine the Height to which may be carried, in a thousand years, the Power of Man over Matter. We may perhaps learn to deprive large Masses of their Gravity, and give them absolute Levity, for the sake of easy transport. Agriculture may diminish its Labour and double its Produce; all Diseases may by sure means be prevented or cured, not excepting even that of Old Age, and our lives lengthened at pleasure even beyond the antediluvian Standard. O that moral Science were in as fair a way of Improvement, that Men would cease to be Wolves to one another, and that human Beings would at length learn what they now improperly call Humanity!

—Benjamin Franklin, 1780

Science! true daughter of Old Time thou art!
Who alterest all things with thy peering eyes.
Why preyest thou thus upon the poet's heart,
Vulture, whose wings are dull realities?
How should he love thee? or how deem thee wise,
Who wouldst not leave him in his wandering
To seek for treasure in the jewelled skies,
Albeit he soared with an undaunted wing?
Hast thou not dragged Diana from her car,
And driven the Hamadryad from the wood
To seek a shelter in some happier star?
Hast thou not torn the Naiad from her flood,
The Elfin from the green grass, and from me
The summer dream beneath the tamarind tree?

SONNET—TO SCIENCE
—Edgar Allan Poe, 1829

INTRODUCTION: WHO OWNS SCIENCE?

EVERY TWO YEARS THE NATIONAL SCIENCE BOARD COMPARES America's science and technology performance to the rest of the world. The 2010 report contained several nuggets of information from polling about Americans' attitudes toward scientific breakthroughs: 68 percent of Americans said that the benefits of scientific research strongly outweigh the harmful results, and only 10 percent said that harms outweigh the benefits. Other surveys confirm these results. As the Pew Research Center reported based on its own 2009 polling:

> Americans like science. Overwhelming majorities say that science has had a positive effect on society and that science has made life easier for most people. Most also say that government investments in science, as well as engineering and technology, pay off in the long run. And scientists are very highly rated compared with members of other professions: Only members of the military and teachers are more likely to be viewed as contributing a lot to society's well-being.

But there is also an undercurrent of unease. In spite of its generally upbeat findings, the National Science Board also found that nearly half of Americans believe that "science makes our way of life change too fast." And it seems that the authors of the National Science Board's report excluded some survey results from the final draft, results showing that Americans are much less likely than the rest of the world to accept evolutionary theory and the big bang explanation of the origins of the universe. The Board said the less encouraging data were excluded from the final draft because they were flawed, but a White House spokesman criticized the omission: "The Administration counts on the National Science Board to provide the fairest and most complete reporting of the facts they track." And a science literacy researcher said that the board's decision reflected "the religious right's point of view."

In America, even surveys about scientific controversy can become matters of controversy.

Future Shock Redux

But no particular group, right or left or somewhere else, is immune from the sense that change is accelerating at an ever faster pace with each passing year. The experience of too-rapid change, whether trivial or profound, is a characteristic of modernity. Information technologies are perhaps the sentinel sources and examples of what Alvin Toffler called "future shock" in 1970, right around the time that a young Bill Gates programmed a GE computer his school purchased using proceeds from a rummage sale. Information scientists cite Moore's law, the idea that computing capacity doubles every two years. We've all experienced the anxiety, frustration, and even resentment that accompanies the introduction of a new version of a software product on which we depend, or the realization that people younger than ourselves have adopted a new technology that makes the pace and style of their lives seem very different from our own.

Reservations about rapid technological change are widely shared regardless of political party or philosophy. In America the tension between approval of science and worry about the rapid changes it can bring bubbles up in special ways when moral or cultural choices seem to be involved. We've seen this tension play out time and again in our seemingly endless controversies about the teaching of evolution, reproductive rights, the moral status of the human embryo, the origins of the universe, and nearly all the issues of science that relate to human values.

Sensitivities about science are understandable. People rightly feel that high stakes are involved when science pushes familiar boundaries, and most of all when it seems that our customary and largely workable moral framework is being challenged. Americans seem especially touchy about such challenges. Ours is in many ways a deeply conservative country where the vast majority (generally around ninety percent) consistently report that they believe in God. The prominence of faith among Americans becomes even more striking when compared with modern western Europe, the historic source of America's core Enlightenment values of rationality and science. There, the proportion of believers is around fifty percent. Americans admire science but also treasure traditional values, which are in some ways threatened more by science than any other institution; our attitudes tend to assemble at the extremes. In this sense, America is both the principal product and the main stage for the ongoing drama of the Enlightenment. Here are these universal values of truth, freedom, and equality founded on reason rather than the authority of a church or sovereign rulers. But is reason enough, or does it threaten those very values?

The ever-quickening pace of discovery in biology is an especially volatile source of "wedge" issues in our politics because it puts into question our familiar values about life itself. These questions are particularly clear when human dignity seems to be threatened, as critics charge is the case with embryo-related research. In 2005 the Genetics and Public Policy Center found that three-quarters of Americans opposed human embryo cloning for research. A 2008 survey sponsored by the conservative Ethics and Public Policy Center found that when the question of embryonic stem cell research is put in terms of curing disease most favored the research, but when described as destroying embryos a small majority opposed it. Five polls by the Pew Forum on Religion and Public Life from 2004 to 2007 found that a majority agreed that it was "more important to continue stem cell research that might produce new medical cures than to avoid destroying the human embryos used in the research."

These results suggest how conflicted Americans are about basic questions of science and ethics. This is nothing new; deep-seated worries about science that are as old as the Enlightenment itself have been poured into bottles made new by the experiences of the twentieth century. The sociologist John Evans has found that conservative Protestant religious groups in the United States do not reject science per se. Rather, they "are opposed to scientists' influence in public affairs not because they do not agree with their methods, but for moral reasons. . . . [T]he relationship between religious persons and science is far more subtle than the dominant assumption of religious opposition to science due to a total rejection of scientific methodology." The problem is not mistrust of science so much as it is mistrust of scientists.

Biopolitics, Old and New

Biopolitics refers to the ways that society attempts to gain control over the power of the life sciences. Although ideas about the role of biology in politics may be found at the earliest stages of Western philosophy, biopolitics promises to become far more prominent as the power of the modern life sciences becomes ever more obvious. The old politics of biology operated in the dark about the underlying mechanisms in question. The new politics of biology arise in the midst of rapidly growing understanding of basic life processes, with seemingly limitless opportunities to direct individual and social change. Simply put, in the modern politics of biology the stakes are about as big as they can get.

The modern abortion controversy has elements of both biology in politics and the politics of biology, especially as it has been a recurrent theme in the United States since the 1970s. As an example of biology in politics, the positions taken by pro-life and pro-choice forces have served as organizing principles. In an example of the politics of biology, each side attempts to manage the power behind the decision to continue a pregnancy or not. But the binary simplicity of the abortion decision itself (i.e., to abort or not) and the relative straightforwardness of the positions one may take on this issue in its strictly political sense (pro-life or pro-choice) are being vastly outstripped by the scenarios forced upon us by the new biology. As biological knowledge grows and as its applications become available, far more complicated and subtle new issues will emerge that can be brought under the heading of biopolitics, the new politics of biology.

The term *biopolitics* was popularized by the French philosopher and historian Michel Foucault. He saw biopolitics as an instance of biopower, the management of bodies and the collections of bodies that we call populations. Key to understanding this idea of biopower is suspending the standard modern tendency to think of the state as the main or even the principle locus of power. Rather, as the philosopher Jason

Robert has observed, Foucault's focus is on those powers among people who have certain key positions in the knowledge economy.

> [B]ureaucrats, administrators, public health nurses, teachers, physicians, genetic counselors, psychotherapists, statisticians, economists. The political government of individuals is effected through special competence and disciplinary credentials. . . . Foucault documents a new power over life, distinct from the right of the sovereign.

Classically, power over bodies and populations was expressed through the idea of governmentality. Not limited to state power, and subsuming even sovereign authority, societal institutions created since the Enlightenment guide conduct in both personal and public matters. As the requirements for a rationale for such arrangements intensified through the emergence of the liberal state, so has the role of expertise, such as the specialized knowledge of the statesman and the scientist. Competition and conflict among parties contending for control over both the actual results and the symbolism of biology have also intensified since the Enlightenment. As Foucault describes it, biopolitics "is the endeavor, begun in the eighteenth century, to rationalize problems presented to governmental practice by the phenomena of a group of living human beings constituted as a population: health, sanitation, birthrate, longevity, race."

Prior to the Enlightenment, Foucault argued, the sovereign exercised supreme power over life with the threat of death. With the rise of rationality as a criterion of acceptable sovereignty, the modern state asserts control not merely over life and death but also over ways of living. The justification for the exercise of this biopower is the need to regulate labor, punishment, public health, reproduction, and various other core cultural habits for the sake of social well-being. Biotechnology may now be added to the list. In the words of another writer on Foucault and biopower, "[g]enetic engineering and genetic-based pharmaceuticals, among other biotechnological pursuits, share an approach aimed at identifying and engineering what are seen as the most basic components of life."

Foucault's idea of biopolitics must be brought up to date. In the more than quarter century since his death, we have entered what has been called the "biological century." If that description is accurate, what does it mean for our politics? The anthropologist Paul Rabinow puts it well: "My educated guess is that the new genetics will prove to be a greater force for reshaping society and life than was the revolution in physics, because it will be embedded throughout the social fabric at the microlevel by a variety of biopolitical practices and discourses."

In the early twenty-first century, we are crossing the threshold to a new biopolitics. Rather than concerning itself with control over bodies and populations per se, the new biopolitics has to do with control over the tissues, systems, and information that are the basis and manifestation of life in its various forms. This new biopolitics is vastly more subtle and, in important ways, potentially more powerful than familiar political struggles over biology, like those having to do with the ability to terminate a pregnancy, or certain clumsy forms of eugenics, and there are already many more protagonists in biopolitics than in the past. Whether the new biology today actually achieves the Promethean power that is often touted, the symbolism alone invites struggles for control. The government, private sector, and scientific community all risk a grave loss of confidence in their ability to manage the emerging forces that the new biology seems poised to let loose. Even if only some of the predictions bear fruit, the new biology will challenge everything in its path, including our understanding of ourselves as living creatures, the ways we live, our relationship to the world, our social arrangements and values, and our political systems.

Science/Technology/Invention/Innovation

The new biopolitics has taken shape just as two venerable distinctions are, in some respects, collapsing. Technology has been around since at least the beginning of agriculture and on some accounts even extends back to the tools and weapons used by hunter-gatherers. Plato wondered how it was possible for mortals to have knowledge of craft or *techne*. But science is a latecomer. One difference between science-based and nonscience-based technology is that scientific theories often have surprising implications that even their pioneers don't anticipate. A classic example is the fact that Albert Einstein had to be persuaded by Leó Szilárd that the atomic bomb was a practical possibility, partly in light of Einstein's own special theory of relativity, so that Einstein would lend his prestige to a letter alerting Franklin Delano Roosevelt of the potential for a weapon holding massive destructive capacity.

The development of science-based technology is remarkably recent, accelerating only toward the end of the nineteenth century with specific, crafted applications of ideas drawn from the emerging explanatory and demonstrable theories, especially in biology. And of course it is still possible to engage in technical manipulations of the world without paying attention to any underlying theory, so science and technology will never be identical. But there is every reason to believe that the convergence between science and technology will go on indefinitely. For a time the idea of starting with a scientific theory as a way to solve a practical problem was so novel that the term "applied science" was used. But so much technology is now science-based, as in the development of new microprocessors, that what used to be called applied science is often virtually synonymous with technology.

To appreciate the traditional relationship between technology and invention, let us take the example of Thomas Edison. He was both a non-science-based technologist and an inventor. The incandescent lightbulb was built on a diverse array of gradually improved materials and owed its origins only very indirectly to electrical theory (of which another great American inventor, Benjamin Franklin, was an early investigator). Alexander Graham Bell was another to whom the term technologist/inventor applies. Both Edison and Bell were brilliant craftsmen who addressed a technical problem. But neither was an innovator. Innovation, in the words of the historian Harold Evans, is more than inventing a new technology. It involves "a universal application of the solution by whatever means. . . . Invention without innovation is a pastime." Universal application is a matter of dissemination, of moving an ingenious solution out into the world. In that sense, the telephone as an innovation is owed to someone who is hardly a household name: Theodore Vail, the president of AT&T. His vision and organizational genius turned Alexander Graham Bell's technology into a national telephone system through the merger of Western Electric and the Bell Company.

The distinction between invention and innovation is more formidable, because it is usually still true that what works in a lab could be prohibitively expensive to disseminate or might not be publicly acceptable. But in some cases, the Internet has virtually (the pun is coincidental but fortunate) eliminated the costs of innovation. The Pentagon's invention of the Internet in the 1960s created the opportunity for innovators like Tim Berners-Lee to develop the World Wide Web. Reminiscent of AT&T's Theodore Vail, who married two entities to produce his communications system, Berners-Lee joined hypertext to the Internet to produce the Web. Today, thanks to that fantastic resource, it is possible to invent an iPhone application and disseminate it almost immediately with hardly any capital requirements on the part of the inventor/innovator. Unlike the case with energy, where the costs of moving from invention to innovation are notoriously high, where the key product is information the moment of invention is also the moment of innovation. With little notice, a similar convergence of invention and innovation is happening in laboratory biology, as genetic sequences can now be e-mailed to labs around the world and chromosomes reconstructed from the biochemical data. In this sense as well, ease and immediacy of scientific communication are giving the scientific community leverage as a new invisible college and are also constituting a global force, a world polity of instantly shareable knowledge and innovation.

Biopolitical Organizing

It is no accident that biopolitics is coming into its own just as knowledge of basic biological mechanisms is beginning to present opportunities for remarkable medical interventions. Previously, the concrete power of biology and contributions of basic biological knowledge to human health had been a matter of debate. The extension of the human life span in the developed world since 1900 has until recently been almost entirely attributable to improvements in public health, particularly the eradication of infectious disease through improvements in water supplies and personal hygiene. As a wag once observed, whomever invented underwear was perhaps the greatest contributor to public health of all time. However, it seems that in recent years a growing portion of the developed world's increased average life span is due to medical interventions, especially in the elderly. As more is learned about gene expression and cellular processes, these interventions can take place earlier in life, resulting in less suffering through disease prevention and perhaps still longer lifetimes. If longer lives are also lives of high quality, the benefits for human flourishing could be vast, but the power that underlies these improvements will, like all sources of power, be a matter of contention. In the midst of these struggles for control, both the legitimacy of the life sciences as governable and trust in the goals and practices of scientists themselves will be at risk.

In the past few years a handful of thinkers and activists have explicitly and implicitly recognized the new biology as a new way of organizing around political values. The questions raised by all sides in biopolitical debates are of ultimate importance to the way we see ourselves as a society and so, unlike many political questions, the usual ideological labels are poor predictors of policy positions. The anti-genetic-engineering crusader Jeremy Rifkin was perhaps the first political organizer to notice that anxieties about the implications of modern biology cut across the familiar left-right political spectrum.

> The current debate over embryo stem cell research, as well as the debates over patents on life, genetically modified foods, designer babies, and other biotech issues, is beginning to reshape the whole political landscape in ways no one could have imagined just a few years ago.
>
> Although reluctant to acknowledge it, both social conservatives and left activists are beginning to find common ground on a range of biotech-related concerns. . . . The threads that unite these two groups are their belief in and commitment to the intrinsic value of life and their growing opposition to what they perceive as a purely utilitarian perspective on biotech matters being extolled by scientists, politicians and market libertarians.

These issues have already begun to make for strange political bedfellows. Some on the left oppose these changes as further threats to human equality, while some on the right worry about the implications for human dignity. Alliances of convenience will develop as people with differing political sympathies make common cause when these issues arise. All but a few libertarians, radical technophiles, and probusiness capitalists have at least some reservations about these kinds of developments. As Rifkin notes, "[i]f the convergence [between social conservatives and left activists] continues to pick up momentum, conventional politics could be torn asunder in the biotech era."

In a telling foretaste of the new biopolitical alliances to come, consider the shortage of organs for transplant. The medical and bioethical establishments favor altruistic kidney donation. This has been the mainstream view ever since transplants from living donors became feasible. But there is not nearly enough supply to satisfy demand, leaving thousands to die of kidney disease each year. Recently, however, a prominent conservative intellectual has joined forces with a well-known pro-choice advocate to challenge the public policy that prohibits compensating organ donors. Meanwhile, most cultural conservatives and social liberals worry about the moral and social implications of paying for organs, even though lives could be saved.

The philosophical intersections that grow out of the new biopolitics can be mapped. Mainstream bioprogressives align with traditional business conservatives in favoring private enterprise. Bioprogressives on the left emphasize regulation, equality, and the common good, while bioprogressives on the right are often of a libertarian cast, emphasizing free enterprise as the most reliable source of innovation. Bioconservatives include both religious traditionalists, mainly Christian, and secular neoconservatives who do not appeal directly to religion but rather to certain traditional religious values in their critique of science, which they regard as a threat to human dignity and moral equality. Some appeal to a core concept of human nature itself. "Green" progressives harbor deep doubts about the implications of science for social justice, often striking a distinctly bioconservative note. A small but growing and vocal philosophical movement, transhumanism or "Humanity+", largely embraces technological change as promoting, rather than jeopardizing, the very values cherished by bioconservatives. In spite of some important dangers, transhumanists see the possibilities for enhancing human nature, while bioconservatives regard human nature as too precious and fragile to withstand manipulation.

Quite different understandings of the history and implications of science and technology and the ability of human beings to adapt to moral challenges are at the core of these philosophical differences. Perhaps with more dialogue about the core differences, the policy disagreements may be ameliorated. After all, if many on the left harbor doubts about science, they have nonetheless not been driven into the arms of social conservatives. Nor are many social conservatives as negative about science as some rhetoric would suggest. We might hold out the hope that all sides could be convinced that science, within carefully negotiated limits, can enhance and enrich the quality of our spiritual as well as our material lives. This is, in essence, the mainstream liberal and progressive view. Yet I think important differences among these novel biopolitical alliances will remain, differences rooted in quite different understandings of the relationship between scientific ways of thinking and human rights, as well as lingering and characteristically post-Enlightenment reservations about the trustworthiness of scientists themselves.

In a way, of course, these political realignments are only new ways of shuffling an old deck. Like generals, political organizers are good at fighting the last war. For those perceptive enough to identify them, however, the new biopolitics also creates opportunities for novel forms of organization and innovative social movements. As is true of the new biopolitics in general, there are already clear signals of what is to come. Take the case of advocacy concerning the needs of persons with certain diseases, disorders, or disabilities. Polio sufferers and their families, persons in wheelchairs, cancer patients, and others have come to be powerful interest groups, securing funding and publicity for massive public health programs, accessibility measures like curb cuts and ramps, and government support of cutting-edge research programs. It is now common to speak of "disease communities," a twentieth-century form of affiliation and self- and mutual identification. Those advocating on behalf of research funding for diseases that are too uncommon to have much political clout on their own have organized into rare disease coalitions. Perhaps the most vivid examples of the legislative possibilities of these efforts are the long-term growth of the National Institutes of Health budget and the passage of the Rehabilitation Act in 1973.

One group that has explored the implications of this new kind of movement is the Little People of America (LPA). Since its founding in 1957, the group has scored impressive gains in both concrete public policies and intangible public attitudes toward those of short stature. Increasingly, members find themselves at the interface of prevalent conceptions of the "normal body" and the growing number of ways to use biotechnology on behalf of a chosen bodily identity. At least some couples who both have achondroplasia, a genetic anomaly that causes short stature, would prefer to have children with the same condition. They want their children to feel fully part of the culture of their community, as they define it. Similarly, there is long-standing division among people with hearing impairments about whether cochlear implants are culturally acceptable or reinforce a stigmatizing notion of disability.

Short stature and hearing impairments are physical conditions that have opened the door to political organization, a sense of community, and even a redefinition of culture. Still more profoundly, genetic knowledge is creating a novel sense of deep kinship that is founded on genetic identity itself. As Rabinow puts it, "There already are, for example, neurofibromatosis groups who meet to share their experiences, lobby for their disease, educate their children, redo their home environment, and so on. . . . [I]t is not hard to imagine groups formed around the chromosome 17, locus 16,256, site 654,376 allele variant with a guanine substitution." Not only does modern genetics create a sense of community among those with certain conditions, it makes it possible for people to select for children with the same conditions. Some conditions will be physically manifest, some will not, but in either case they will change the ways that people view their shared interests. In other words, politics will increasingly become biopolitics.

Though of course moral questions about reproduction stand in the background of much of our biopolitics, we may be confident that these questions will themselves be transformed by events we cannot anticipate, in both science and public affairs. Some of the emerging topics I will discuss are directly related to the politics of reproduction, others to the ways that we die, and still others to the remarkable prospects for new directions in health care, in knowledge about our biological nature, and for the enhancement of "natural" capacities in ourselves or our children. Formerly clear lines will be blurred; inquiry has a familiar and sometimes annoying tendency to upset comfortable ways of thinking. Although the results will almost certainly not conform to our most confident predictions, both our reasonable expectations and the surprises in store will force reexamination of the ways we think of ourselves as individuals and about the ways we arrange to live together.

America/Future/Progress/Science

Consider, finally, the constellation of these ideas. The American dream is inextricably tied to the vision of a future of progress founded on science. To the extent those prospects are threatened, so is the dream and so, therefore, is American's civic religion, its *raison d'être*. That is why, especially in America, new biological knowledge aggravates our cultural tensions, stimulates further debate about the legacy of science in our sense of national purpose, and challenges our political system and the values implicit in our public life. Over the horizon, there are still other plausible developments that, if they come to pass, could fundamentally transform human society. How will America define itself in the century of biology? If politics is, as I believe it is, ultimately the only alternative to violence, these matters are worthy of the best politics we can muster.

CHAPTER ONE
SCIENCE IN AMERICA

THE NEW POLITICS OF BIOLOGY IS UNFOLDING AGAINST A COMPLEX cultural and historical background of scientific and technological innovation. The history of science in America reveals certain themes that are already at play in the new biopolitics: the tension between innovation and tradition, the influence of the Enlightenment in America's self-understanding, the importance of science for American power and prosperity, the relationship between progress and science, the role of government in sponsoring technological innovation and advancing a vision of science, and the development of a typically American "pragmatic" philosophy. What does America's experience with science suggest about its ability to muster the imagination and energy for a biotechnological future? How do the life sciences fit into America's civic narrative? And how has that experience laid the groundwork for the new biopolitics?

American Science and the Enlightenment Legacy

If a coherent historical narrative is important in sustaining a sense of national commitment to science, then America has a unique advantage, for in that respect the American founders' philosophical orientation and practical interests laid an unrivaled foundation. The basic liberal idea of universal human rights flowed from Enlightenment values of free speech and tolerance. With few exceptions, the philosophers of the Enlightenment from whom the founders drew inspiration viewed science as the most reliable source of knowledge and the only one that is independent of arbitrary authority. Considering how little time has passed since the seventeenth century, at least in the material sense, the human progress made possible by the Enlightenment and the scientific revolution it simulated has been astonishing. Measured against the scale of the appearance of *Homo sapiens* 200,000 years ago and the advent of agriculture and stable settlements within the past 12,000 or so years, improvements in both the length and quality of life are now so familiar that the rapidity of their appearance is hard for us to appreciate. One rough if reductively actuarial measure is per capita income, which was about $700 a year in 1800 (virtually unchanged for a millennium), and about $6,500 in 2008. In 1800 most people died by the time they were thirty; today life expectancy is about seventy-eight. In spite of global inequalities, malnutrition in the developing world is one-third what it was in 1945. Memories are short; we are so aware of the woes of modernity that we forget the opportunities for human flourishing represented by the Enlightenment.

Demands placed on the state have also drastically changed since the Middle Ages, as individuals came to see themselves as bearers of rights and with reasonable expectations as members of civil society. Before the Enlightenment, "the pursuit of happiness" would not have been a criterion of successful government. The pursuit of happiness is a public affair that requires beneficent progress. As the sociologist Richard Nisbett observed, "Few places in the eighteenth century displayed a stronger belief in the philosophy of progress than did the American colonies and, then, the new, infant republic." For the revolutionary generation, science, understood as rational argument and demonstration, was also part of the constellation of ideas that gave the United States special promise; it is not too much to say that America is the only country founded by a group of scientists. No wonder that Americans reflexively and often unconsciously identify their country with the idea of progress, in spite of their reservations about its implications. Neither is it a coincidence that, along with a preoccupation with progress, Americans have also engaged in an intense struggle over the meaning of human rights since the declared inception of an American nation. To say that the criterion of the truth of human equality is its "self-evidence" is to imply that none could rationally deny the right to rebellion when "inalienable rights" are infringed. Even the greatest power in the world, in this case the British Empire, is nothing compared to the power of reasoned, scientific insight, to which no earthly authority has privileged access.

Under the influence of John Locke's moral philosophy, the Declaration of Independence's assertion of self-evident truths was presented as a simple acknowledgment of their axiomatic status. Axioms are part of a web of logically implicated statements. In empirical science these statements also include propositions that are subject to experimental confirmation. Taken together, this set of true statements may constitute a system that is unwelcome to established authorities. As the American founders understood, and as they themselves implemented in their insurrectionary practice, science is subversive. The subversive nature of science does not directly actuate political activity . . . does not directly actuate political activity. Though the combination of intuited laws and experimental method has profoundly influenced social change, the ultimate measure of science lies in the ways we moderns perceive, understand, and act upon the world. Science is no respecter of the preferences of the powerful. It challenges prejudices, obscures boundaries, and undermines familiar categories. It threatens comforting and stultifying dogma. Although scientific ways of thought inspired the radically innovative American idea of self-government, the founders did not believe that scientists should govern —a suggestion Francis Bacon seems to have been the first to make, and one that has ever since unnerved many. It is rather that an approach founded on trial and error, observation and calculation, as the only alternatives to mere authority, whether its source is metaphysical, theological, or political, should prevail.

The Founders as Scientists

One can draw a line of descent from Isaac Newton, the greatest scientist of his day, to Benjamin Franklin, the greatest scientist of his. By carefully observing and mathematically calculating the effects of gravity, Newton showed that one could predict the motions of the heavenly bodies. Reading *Principia*, John Locke was struck that generations of philosophers had been so preoccupied with their own comprehensive metaphysical systems that they failed to be open to the lessons of experience. Similarly, it seemed to Locke that abstract and interminable arguments about such "problems" as whether human beings are naturally free lead political philosophy down a blind alley. The point is rather that people are at liberty to do what they decide to do and that government should focus on conduct. The people, too, should respect state sovereignty only insofar as the actions of the state redound to the protection of their rights, a social contract.

Locke appreciated that Newton's scientific approach gave the lie to the notion that one had to be a philosopher or priest or king to know the nature of things. Rather, political liberalism is founded on the fact that sovereign authority has at most an incidental relationship to the truth, that insight into the nature of things is independent of power or social status. Enlightenment epistemology values a worldview's correspondence with reality, proven through demonstration rather than its internal coherence alone; while the latter is considered a logical requirement and an aesthetic virtue, coherence in itself is not a criterion of knowledge. Ancient cosmologists went to great lengths to ensure the internal coherence of their Earth-centered universe, but elegance is not enough. Even the presumptive axiomatic statements that are the lawlike generalizations at the core of a system of statements about the world are themselves subject to disconfirmation by experiment.

Thomas Jefferson was famously preoccupied with both astronomy through Newton and political liberalism via Locke. Mathematical reason and empirical observation were not to be set off against each other as in pre-Newtonian philosophy, but viewed as complementary and, so much as possible, given the limits of contemporary knowledge and opportunities for experiment, integrated. Thus, the pragmatic and progressive strains in American thought and Americans' self-understanding reach back to the founders and to their inspiration in Enlightenment and classical liberal figures like Locke and John Stuart Mill. The revolutionary generation largely accepted the Enlightenment's confidence in progress, the idea that through empirically based administration of public affairs (known today as the policy process) it was possible to improve the human condition. In principle, these improvements were limitless. Yet the founders' experience with unreasonable established authority in matters of taxation and personal religious conviction led them to construct a system in which state authority was constrained: the famous "checks and balances" of the three branches of government.

The founders reached a consensus on a republican form of government that was finally buttressed not by a democracy per se, but a constitution. An underlying motive for this cautious attitude toward democracy was anxiety about the fragility of social arrangements, of civilized mores. The dislocations associated with technological developments have long been a source of concern among social conservatives, as have any sudden eruptions in the political fabric. The father of Anglo-American conservatism, Edmund Burke, was famously and quite rightly alarmed by the chaos, criminality, and corruption unleashed by the French Revolution; his visceral response was to invoke fealty to tradition. "The very idea of the fabrication of a new government," Burke wrote, "is enough to fill us with disgust and horror. We wished at the period of the Revolution, and do now wish, to derive all we possess as an inheritance from our forefathers." Shocked by the betrayals and violence of the Terror, Burke was by contrast a brave and often lonely supporter of the American rebellion, perhaps because the colonial governments were not "new" but longstanding, finally becoming irreparably alienated from a distant monarchy while still retaining English roots.

In sustaining a delicate balance of idealism and realism about human affairs, the American founders adopted an attitude we might today call ironic, but might be better called fallibilism. This term was used by an American philosopher a century after the revolutionary period to describe his philosophy of science: all statements about the world may be proven wrong. This American "pragmatism" anticipated the later twentieth-century positivist view, originating on the continent, that all empirical statements must be falsifiable. The founders were only too well-aware that events could surely have falsified their theory that popular self-government is possible. It was thus perspicacious of Jefferson to describe their project as the American experiment. As a man of science, he well knew that he and his colleagues could be wrong. Franklin himself, while acknowledging his doubts about the perfection of what would be the final draft of the Constitution, urged each of his colleagues to, "on this occasion, doubt a little of his own Infallibility, and to make manifest our Unanimity, put his name to this instrument."

The founders' ironic, fallibilist vision of human beings and human society, and even of their own constitutional design, therefore led them to favor a popular government but one that was constrained by certain institutional arrangements. Their version of an Enlightenment governing philosophy deserves to be thought of as pragmatic in at least two senses: first, that it would be put to the test in experience; and second, they viewed human nature as improvable but inevitably flawed. The latter sense of the founders' pragmatism led them to favor careful limits on state power. Key members of the revolutionary leadership were pragmatic experimentalists who were in a position to glimpse the coming emergence of the modern idea of science and the way that a nation could exploit the growth of knowledge. Jefferson provided perhaps the most obvious and, over the long term, concretely influential example with his authorship of the patent clause "to promote the Progress of Science and useful Arts," which provided incentives for inventors that largely remain in place today and are still exceptionally generous. Jefferson saw intellectual property as providing both a magnet for creative individuals and a source of improvements for the common good.

We have recently come to know Jefferson as a more flawed, complex, and interesting person than his former misty image as the sage of Monticello. Not debatable is his intellectual brilliance. While Secretary of State, Jefferson personally reviewed each one of the first set of patent applications, later confessing his insecurity as an examiner ill-prepared to take on fields that were already showing signs of specialization. His preoccupation with the design of immensely clever gadgets was a traditional technologist's approach uninformed by theory. One senses in Jefferson a deep curiosity about underlying processes that were invisible to him, but the time was not yet fully ripe for applied science. Similarly, Benjamin Franklin's interest in electricity and metallurgy manifested themselves in his invention of lightning rods and circulating stoves. To say that Franklin was obsessed with the value of experiment and observation is no exaggeration; he even conducted experiments on temperature and atmosphere during his frequent perilous ambassadorial voyages across the Atlantic.

Although the intellectual interests of Jefferson and Franklin were perhaps the most highly developed of their peers, others in the revolutionary pantheon showed a keen appreciation for natural philosophy. As a student at King's College, now Columbia University, Alexander Hamilton was intent on a medical career and attended all the lectures on natural philosophy that he could. John Adams, John Hancock, and James Bowdoin founded the American Academy of Arts and Sciences in Boston. Jefferson and Hamilton, otherwise bitter rivals, made complementary contributions to the innovative foundations of the new country: Jefferson through the patent statute, Hamilton by laying the foundations for history's most successful capitalist economy. By way of the wildly popular pamphlet *Common Sense*, Thomas Paine was not only the most effective propagandist of the American Revolution, he also closely followed current scientific breakthroughs. In *The Age of Reason*, he declaimed on the size of the earth, the nature of the planetary system, and the scale of the universe. Paine theorized that there must be millions of worlds like ours millions of miles apart.

With scientists among those at the helm of the American insurrection, hopes for the new nation among educated classes on both sides of the Atlantic were high. In 1794 the ubiquitous British chemist Joseph Priestley told the American Philosophical Society in Philadelphia, "I am confident . . . from what I have already seen in the spirit of this country, that it will soon appear that Republican governments, in which every obstruction is removed to the exertion of all kinds of talent, will be far more favourable to science, and the arts, than any monarchical government has been." It is fair to say that no nation has ever been founded by people who were more oriented toward the pursuit and propagation of knowledge than the United States.

Moral Asylums

Perhaps the most obvious examples of the concrete application of early America's knowledge society were its institutions of social control. Alexis de Tocqueville arrived in 1831 to examine the way a new democracy dealt with its deviants, more than a century before Michel Foucault called attention to biopower as the discipline of bodies and populations. Tocqueville's observations of life in America, published in 1835, have long been objects of fascination in both America and France, as he quickly saw far more to report about than prisons and penitentiaries. Tocqueville identified one of the points of tension between tradition and Enlightenment that continues to flavor American biopolitics:

> Upon my arrival in the United States, the religious aspect of the country was the first thing that struck my attention; and the longer I stayed there, the more did I perceive the great political conse-

> quences resulting from this state of things, to which I was unaccustomed. In France I had almost always seen the spirit of religion and the spirit of freedom pursuing courses diametrically opposed to each other; but in America I found that they were intimately united, and that they reigned in common over the same country.

Had Tocqueville stayed a bit longer, he would have seen the rise of an American institutional connection between religion and freedom, one that also drew on the values of the French Revolution. In 1797, a Parisian hospital physician dramatically unchained the male mental patients in his charge. He allowed the patients to walk around the hospital grounds and introduced light and air into their quarters. The ideas behind an empirically based "moral treatment" of the insane took hold in Britain and the United States as well. The devoutly Christian Dorothea Dix led a movement that resulted in the establishment of several dozen mental hospitals according to architectural arrangements that were supposed to help bring order to the minds of the "distracted lunatics." Underlying the design of these asylums was the epistemology developed by Locke, John Stuart Mill, and David Hume, which proposed that all ideas were imposed by impressions on the mind from sensory experience. Even the location of ducts and pipes was obsessively specified in the blueprints of some of these institutions. Moral treatment to reorder the minds of the insane inmates extended beyond physical arrangements to social structures intended to develop a healthy mental outlook; in some asylums the superintendent and his wife would hold great formal balls for the residents and function as substitute parents in a setting designed on the model of a settled and secure family life.

After the Civil War, physicians questioned the radically empirical assumptions of moral treatment, noting that many inmates obviously had physical or organic problems that were associated with their mental conditions. By 1900, moral treatment asylums, underfunded and overwhelmed by the social ills of industrialization and immigration, deteriorated into the mass mental hospitals later known as "snake pits." They were finally subject to deinstitutionalization policies in the 1970s. Deinstitutionalization took place amid a philosophy that was increasingly hostile to psychiatric authority. One of the critics was Foucault himself, who regarded moral treatment as just another form of subjugation of bodies, as did the antipsychiatry movement that grew out of the 1960s. Yet for all its limitations, moral treatment was a noble effort to give decent and respectful asylum to those who would have otherwise been destitute and abused. It was a vivid demonstration that empiricism can promote humane social goals.

The "Internal Improvements" Debate

Moral treatment asylums were experiments in applying the empiricist psychology of the day but were not in themselves laboratories of innovation. At the level of scientific research and technological innovation, how well did the country exploit opportunities prior to the Civil War? There were several modest postcolonial organizations, including Franklin's American Philosophical Society, the American Academy of Arts and Sciences, and the National Institute for the Promotion of Science. But in the first half of the nineteenth century, formal government engagement in innovation was lacking. At least one of the founders advocated an aggressive role for government in industrial and science policy at the nation's inception. In his *Report on Manufactures*, Alexander Hamilton proposed a plan for "[t]he encouragement of new inventions and discoveries at home, and the introduction into the United States of such as may have been made in other countries; particularly those which relate to machinery." Hamilton's idea of "encouragement" included prizes as well as patents, and what were then known as "premiums" for innovation, similar to today's tax credits for research and development. As General Washington's aide during the Revolutionary War, he was well acquainted with the need for the best possible weaponry and hence the need for cutting-edge inventive capacity. When new products entered the marketplace, Hamilton also believed in "judicious regulations for the inspection of manufactured commodities." Hamilton's vision was thwarted by a suspicious Southern congressional delegation and critics of "mercantilism," only to be reintroduced by President Lincoln and, finally, enshrined in the Progressive Era and our modern national policies.

Sectional loyalties discouraged federal investment in science and modernization for fear of infringing on the autonomy of the states. The controversy about federal authority over "internal improvements" was one manifestation of the conflict that later erupted into the Civil War.

How much power did the Constitution give the central government to influence economic developments in the states? To Southern congressmen especially, the idea that the federal government would insert itself in local improvements seemed a power grab. An early proposal for a national university to be located in Washington, though apparently a dream of George Washington, was caught in this debate. Following in Hamilton's spirit, a few pre–Civil War congressmen, Henry Clay among them, argued that the United States needed to innovate in order to compete in international trade, especially to prevent the dumping of goods on the American market.

There were other cultural differences at work, as some members of Congress also expressed skepticism about providing gentlemen of leisure with support for experimental work. Did speculative work have a place in a new nation busy with the practical problems of creating new civic institutions? In *Democracy in America*, Alexis de Tocqueville was the first to articulate a stereotype of Americans' attitude toward theory:

> In America the purely practical part of science is admirably understood, and careful attention is paid to the theoretical portion which is immediately requisite to application. On this head the Americans always display a clear, free, original, and inventive power of mind. But hardly anyone in the United States devotes himself to the essentially theoretical and abstract portion of human knowledge.

Before the latter part of the nineteenth century there was much truth to this observation. As the historian Charles Rosenberg wrote in his classic work on the social history of science in America, "Only occasionally and then in circumscribed areas did Americans before 1914 demonstrate a willingness to support systematic investigation." There were exceptions. When scientific work presented direct and tangible advantages for their constituents, congressmen were supportive, as in the case of the coastal survey, which aided in safe navigation and therefore enhanced opportunities for commerce and for the defense of the states.

But where to draw the line on internal improvements? The sixth presidency ran aground partly on this question. John Quincy Adams was an aggressive advocate of modernization who, though narrowly elected, pressed Congress for roads, bridges, canals, a national university, and an observatory. His initiatives may strike us today as visionary, but to the growing group of Andrew Jackson supporters, they were arrogant. Though Adams served only a single term, some of his ambitious innovative proposals, such as the canal system, become reality. Unlike Adams, Jackson did not appreciate a role for the national government in economic development or its implications for national power and prosperity. In his two terms, Jackson focused on such matters as tamping down growing hostilities between slave state officials and abolitionists and displacement of Native Americans from Georgia.

Another famous expression of Jacksonian populism that inadvertently inhibited innovation was the dissolution of the Second National Bank, which had been created after the War of 1812 to handle the nation's financial affairs. Opposition to the bank was due in part to popular suspicion of the concentration of wealth in the hands of a few elites, and in fact there was ample corruption in bank management. Proponents of the bank regarded public opposition as a case of anti-capitalist conspiracy theory and a failure to recognize the need for a ready supply of capital to finance innovation. But Jackson had no taste for mere bank reform and pressed instead for a revocation of its charter, which now seems to have been a mistake. When Jackson distributed funds from the dissolved bank into smaller state banks and other local projects, the result was a temporary boom followed by inflated bank notes, debt, and a severe depression. Lessons learned from this experience were incorporated into the modern, decentralized Federal Reserve Bank.

A by-product of the pre–Civil War financial crisis spurred by the end of the Second Bank was a poor economic environment for investment in new ideas. Through no effort of Jackson, the end of his administration happened to coincide with the advent of one of history's great transformative technologies, the telegraph. In 1838, Samuel Morse showed that his electromagnetic design could be combined with his dot-and-dash code to produce a practical device for long-distance communication. But there was no immediate government support for the transition from the invention of the telegraph to the societal diffusion needed to make it a true innovation, nor was there a source of private capital for such a speculative venture. Although Congress was favorable in principle to the concept of the telegraph, it was five years before it funded an experimental telegraph line from Washington to Baltimore. A program of internal improvements as advocated by Henry Clay and

John Quincy Adams might have sped the creation of a telegraph system. As it was, the transcontinental telegraph system was not in place for another fifteen years, just in time for the Union Army to make the most of the new instantaneous communications for tactical purposes and for the terrible costs of the war to be relayed in frequent dispatches across the continent.

Perhaps the second most important innovation in nineteenth-century America after the telegraph was the transcontinental railroad, delayed by Southern senators' pre–Civil War reservations about internal improvements. The project was also viewed skeptically by those who doubted that any rail line could conquer the Sierra Nevada mountain range. Finally, after securing support from the California legislature and commitments of private investors, the visionary and indefatigable Theodore Judah succeeded in convincing Congress to pass the Pacific Railroad Act in 1862, which included a provision for U.S. thirty-year bonds to finance construction. Like the purchase of Alaska, the railroad was a brilliant investment that could only have been made by a nation committed to federal authority. It also didn't hurt that President Lincoln had seen to his own education on railway technology.

In general, the country's record on governmental investment in innovation before the Civil War was mixed at best. The slow response to the importance of the telegraph was not simply a failure of imagination and testimony to the simmering national crisis, but also evidence that there was no clear system of cooperation between government, academia, and the private sector. As is so often the case, war provided the impetus for a more organized approach to science and technology. Lincoln created the National Academy of Sciences in 1863 to evaluate proposed inventions that could aid the war effort. Its Act of Incorporation provides that "the Academy shall, whenever called upon by any department of the Government, investigate, examine, experiment, and report upon any subject of science or art. . . ." Before that, Lincoln persuaded Congress to pass the Morrill Act establishing land-grant colleges "to teach agriculture and the mechanic arts," but also science, classics, and military tactics. President Buchanan had vetoed an earlier version of the bill. Undoubtedly, these initiatives to strengthen the federal government's hand in encouraging the spread of learning and the innovation it stimulates benefitted from the fact that opposition to federal power from the Southern states was temporarily absent from Congress.

The founding of the National Academy punctuated other cultural trends, including the emergence of the modern meanings of the words "science" and "progress." In the first half of the nineteenth century, inventors and laboratory workers in Europe and America were developing microscopes and improving telescopes to drill both down and up into nature, while stethoscopes began to reveal the body's inner space. Naturalists like Darwin embodied openness to perceiving complex natural relationships without preconceptions. In places like Berlin and Vienna, "scientists" (that word, too, was coming into vogue by mid-century) were learning that electricity was a force in physiology and theorizing about a newly visible world of microbes. Systematic investigations that manipulated variables proved more revealing than mere observation. The possibilities that could emerge from human insight began to seem endless.

Especially in the industrial boom times after the Civil War, moral values were seen as a key consequence of scientific progress. Rosenberg has observed that "[t]he vast majority of nineteenth-century Americans never doubted that human beings had progressed and that this progress—inevitably—subsumed dimensions both moral and material. It was inconceivable to them that the steam engine and morality were not somehow interconnected." Piety, productivity, and, by the end of the nineteenth century, efficiency were all within the same universe of desirable values and consequences of science and progress. So, too, was nationalism.

> To the earnest young advocates of agricultural science in the 1850s, America's peculiar virtues were unquestionable. . . . Though American students might concede that German pure science led the world, their own American countrymen seemed far more skillful and ingenious in the application of science and technology to the improving of man's lot.

Physics, engineering, and chemistry were regnant, and biology was still largely observational rather than experimental, so the great debates about evolution and the origins of life were yet to come. Partly for this reason, through the late nineteenth century conservative religious beliefs were quite compatible with a cohesive moral vision. According to one historian, "Evangelical Protestantism and science were so intellectually compatible in the United States that a naturalist and a minister could easily agree on what they believed about nature."

Nineteenth-century obstacles to investment in science and innovation were financial and political, not moral. Yet it took decades for agriculture experiment stations—which sought to introduce science that would make farming more efficient and competitive—to receive government funds. In 1861, an agriculturalist described his experience in one state: "I have spent the whole vacation dogging at our legislature for money. I have been put off, trifled with, cheated, deceived and humbugged in a great variety of ways till now the session is nearly to a close and yet not one dollar voted." Modern lobbyists should take note. Gradually, though, government support of development was breaking through. The first bill that provided public lands for state colleges was included in the 1862 Morrill Act. Later grants opened the Great Plains to settlement on generous terms. Federal support of agricultural science began in 1887 with the Hatch Act.

American industrialists and their European counterparts were taking capitalism to new levels of innovation and human improvement, resting in part on the cultural conditions that German sociologist and economist Max Weber called the Protestant ethic. The spirit of adventure and possibility was signified in the very physical spaces of America, a great stage that could nurture virtue and improvement. In 1893 historian Frederick Jackson Turner delivered his paper, "The Significance of the Frontier in American History" to the American Historical Association. Whether or not his thesis that the frontier was the key factor in America's uniquely innovative and democratic character was accurate, the idea itself captured the imagination of generations of scholars and the educated public, becoming very nearly a self-fulfilling account. Like Peirce, the philosopher of science, Turner was deeply influenced by evolutionary theory. According to his thesis, the seventeenth-century settlers were in effect forced to take an experimental attitude toward their situation, as the old European ways would not do.

The frontier thesis continues to capture the way Americans would like to see themselves, if not how they actually are. As modern historians have noted, by the time of Turner's speech, the frontier had just about disappeared. Even then there remained the inconvenient truth that the hinterlands were anything but unpopulated when the Europeans arrived; evidently, the natives didn't count for much in practice or in theory. But there is no denying that the idea of the frontier has had a powerful hold on the American mind, as the conquest of the West was widely seen as the very embodiment and execution of the progressive spirit. German intellectuals who settled in St. Louis during the later nineteenth century came to recast their Hegelian sympathies as specifically fulfilled in America, as the World Spirit, like the sun and civilization itself, made its way westward.

Evolutionary Philosophy

At about the same time, naturalists were developing a somewhat more complicated view of the moral implications of human endeavor. Observations and inferences drawn from the origin of the species also led to a deeper understanding of the nature of science. No scientific theory has more upset the nature of order—and the order of nature—than evolution. Daniel Dennett has called evolutionary theory "universal acid," because it eats through all that surrounds it. Most threatening to a conservative worldview, evolution upsets two fundamental and closely related assumptions of virtually all previously recorded worldviews: that order can not emerge from disorder, and that the appearance of design all around us, and especially in the biological world, implies design. Darwin told his friend Joseph Hooker that in reporting his conclusions, he felt as though he had killed someone. The remark was extreme but not inappropriate. We mourn the passing of our cherished illusions. No system devised by human beings has ever challenged more dramatically than science, and no science more than the new biology. The social implications of the burst of the new scientific thinking in the seventeenth century have ever since been a matter of concern for virtually every important thinker.

What Darwin appreciated was that, through some process invisible to him, given enough time, nature could in effect try all sorts of combinations, some of which proved to be successful for whatever the current environmental challenges happened to be. Life finds a way. In this sense, his contribution was not only about biology, but was another demonstration of the power of evidence, reasoning, and acute observation. A later nineteenth-century thinker, an American, was perhaps the first to appreciate the implications of Darwin's work for the logic of science. An obscure figure to all but aficionados and scholars of American philosophy, Charles S. Peirce is the lynchpin in the way that the idea of science itself evolved as part of America's self-understanding. Peirce, a difficult genius who died in penury, published two papers in *Popular Science Monthly* in 1877 and 1878 that remain the most powerful statements by an American on the nature of science and its advantages over all other ways of attempting to know the world. Citing Darwin's "immortal work," Peirce presented a message of knowledge by observation, experimentation, and calculation that was also deeply Newtonian—and appropriately so, as Peirce's father was a Harvard mathematician.

Still, Peirce rejected the notion that nature is random. He insisted that there is order in the universe and, like Hegel, that ideas are not just in our minds but are real and move history. There is, however, also irregularity in the universe. Laws evolve as the universe itself is in a process of growing complexity and diversity; regularity emerges from irregularity. All our ideas, including scientific theories, are rendered more precise through our experience with this unfolding reality. Arguing that the meaning of an idea consists of its practical effects, Peirce inspired America's native (and often misunderstood) philosophical movement, pragmatism. Theory is thus intimately tied to practice and ideas to the actions they call forth. Similarly, later pragmatists like William James and John Dewey argued that mind and body are not different entities but continuous with one another. So therefore are our ideas and our behavior, and thoughts should be understood as dispositions. This action-oriented approach to theories and ideas is one of the distinguishing features of philosophy in America, and it reflects a distinctive American sensibility.

As dispositions to action our ideas are subject to demonstration. They form the basis of and achieve meaning in experimentation. In "The Fixation of Belief," Peirce observed that, as compared with tenacious stubbornness, religious authority, or a priori philosophies, only science tests its ideas against reality. Truth is the relationship between a statement and reality, and though reality cannot ever be fully known, through experimentation we are able to refine our belief-statements and, as a community of inquirers, come closer and closer to a precise account of that which is real. Peirce realized that this account of truth is only accurate if, with an unlimited amount of resources, the scientific community would in principle produce a set of statements that perfectly reflected reality. In fact, we never achieve that state, but in principle, if allowed to run its course, over the long run science does indeed bring us closer to the truth. The same cannot be said for any other way of "fixing beliefs" about the world, like those derived from the authority of metaphysicians, because, however grand their views, they are not subjected to experimental test. There is no way of gaining knowledge, or "fixing belief," of having a good reason to believe what we believe, apart from the sort of publicly accessible inquiry of which science is the most systematic expression. We may choose to believe whatever we like, either because we just want to or have a conscious or unconscious prejudice, or because we have accepted some religious or secular authority's teaching. Without subjecting that belief to the crucible of experience, all these ideas really amount to the same thing. The alternative to experimental confirmation is, in a word, dogma. Dogmatic statements may have many fine qualities. They may be beautiful, inspirational, and convey a kind of wisdom, or at least the impression of wisdom. But they can never be verifiable and self-correcting in the manner of science.

The notion of science as a uniquely self-correcting process is poorly understood and underappreciated. It might be conceptualized as a kind of thought experiment that assigns infinite values to three variables: the number of scientists, the amount of time they have to work, and the resources they have. Given enough scientists, enough time, and enough resources, science will tend to correct for its own errors, as erroneous explanatory theories will be shown up as anomalous, forcing new experiments and improved explanations. The logic of science is that of gradual but never completed self-correction. What is important, though, is that over the long run the ideal is realized, albeit incompletely, and that in spite of this idealized conception of science, the same cannot be said for any other way of comprehending reality.

Implicit in this view is the idea of community, which in this case is the unique and specialized community of expert scientific workers. The notion that improvement can be achieved through social cooperation is characteristic of political liberalism, which metaphysicians need not embrace. The very idea of community as no obstacle to progress, indeed as a condition of progress, is one that in some respects is traceable to the classical predecessors of the Enlightenment liberals. Aristotle, though he lacked a modern view of progress, at least thought that human beings could only be perfected within a certain sort of community, the city-state. To appreciate the importance of community for political liberalism, we might note that this was the main sticking point between Enlightenment liberals and romantics like Rousseau, for whom the natural state of freedom was shackled by the products of human political communities.

Peirce is mainly thought of as a logician, but his indirect influence on American culture went well beyond technical matters of mathematics and logic. Peirce crankily disassociated himself from the influence of his pragmatic account of the meaning of ideas and theories. As transmitted mainly by William James, the inherent ambiguity in Peirce's statement of pragmatism led to a sore point between the two. James interpreted an idea's "effects" or "consequences" far more broadly than Peirce intended. For James, an idea's meaning, or "cash value," was the extent to which it led to a series of perceptions or experiences. For example, James was quite interested in parapsychological phenomena, like the possibility of communicating with the spirits of the deceased. Although he was never convinced that such communication was possible, some of his writings seem to suggest that, in the absence of evidence to the contrary, the "will to believe" in an idea can lead to its own justification, perhaps by giving meaning and purpose to life. Thus, despite his personal struggle with depression, James embodied the resolutely positive "can-do" American persona. (Before his death, James promised his brother Henry that if there was another side he would send him a signal; so far as is known, none was received.)

As a popular Harvard professor and public speaker who always credited and sometimes financially supported the querulous and unemployable Peirce, James was a far superior promoter of his thought. Because of James's compelling writings and personality, the themes at the heart of Peirce's seminal papers have had a profound effect on the way Americans see themselves as "pragmatists," willing to entertain a hypothesis in order to assess its value in practice. In this sense, pragmatism is at the heart of the progressive philosophy of government that came to fruition under Theodore Roosevelt, deeply influencing what we now know as the public policy process.

The Progressive Era

Enhanced sea power, rapid industrialization and urbanization, explosive immigration, new communications technologies, the accumulation of great personal fortunes, and excitement about novel insights into the logic and growth of science were foundation stones of an increasingly self-confident country that saw in its reflection the very essence of progress. By the turn of the twentieth century, the stage was well set for a progressive movement that not only took on the "malefactors of great wealth," but also sought to make government an effective agent of social progress. The systematic collecting of data about the effects of policy and the recruitment of expertise in guiding policy were important elements of the progressive strategy. Both were dependent on the ideas of evidence and inquiry as fundamental progressive values. In 1914 Walter Lippmann wrote, "The scientific spirit is the discipline of democracy, the escape from drift, the outlook of a free man." An early example of the turn toward expertise as informing government action was the Commission on Country Life, appointed by Theodore Roosevelt to issue recommendations on the best pathway to rural development, including electrification. The goal was to bring rural life (which, in spite of the Jeffersonian ideal, was often one of isolation and alienation) up to date: to bring it into line with and enable it to benefit from the technological changes then taking place. At least one aim of this group of urban and progressive activists anticipated by a century the desire of modern environmentalists to instill an environmental ethic in the farm community.

The progressive movement targeted private interests that had amassed enormous wealth and power after the Civil War. Besides the widely perceived dominance of predatory interests, the era was in other ways very much like our own. The notion of equality of opportunity, part of America's popular gospel since the Lincoln administration, was in practice denied to millions, especially factory workers, new urbanites, and immigrants, who lived and worked in deplorable, often Dickensian conditions. There are other provocative parallels to our own time. In 1890, there was a financial crisis, high unemployment, economic uncertainty, anxiety about immigration, terrorist violence, concern about food quality and drug safety, growing alcohol and drug addiction, and increased access to pornography. As Phillip Longman and Ray Boshara note, there were even "sharply rising, politically untouchable, and unsustainable levels of spending on the well-organized elderly. By 1894, a staggering thirty-eight percent of all federal spending went just to pay pensions to Union Civil War veterans and their dependents," compared to forty-two percent for Medicare and Social Security today.

Progressives held at arms length the various forms of socialism sweeping through Europe and attempting to take root in the United States. One of the most influential tracts in the progressive movement was Herbert Croly's *The Promise of American Life*, in which he urged a "new nationalism" that abandoned Jeffersonian individualism and the unconstrained market principles that created an undesirable concentration of wealth. As he wrote, ". . . the national government must step in and discriminate . . . not on behalf of liberty and the special individual, but on behalf of equality and the average man." In that spirit, progressive reforms that began under Theodore Roosevelt and to some degree continued under his cousin, Franklin Delano Roosevelt, addressed virtually all of these social ills and in the process built a modern nation-state.

What ties progressives together, then and now, is a desire for legal measures and the rallying of public support for political candidates who will level the playing field and promote social equality. Progressives continue to be willing to use government to represent the public interest on behalf of the weak and vulnerable. As the earlier progressives had not experienced the vast expansion of government during and after World War II, size was viewed as a corrupting factor for business interests. Supreme Court Justice Louis Brandeis wrote of "the curse of bigness," though not in reference to the apparatus of the state. For this generation, the challenge was not to oppose or abandon government, but to take it back from the plutocrats, for it was believed that only government could level the playing field when markets failed.

Even as it flourished, the progressive movement did not speak with one voice. Adherents were deeply divided on questions of militarism and imperialism. William James, for example, opposed war as an outmoded and immoral form of racial improvement but affirmed progressivism. In 1906 he fired off a famous rejoinder to the president of the United States, who also happened to be an alumnus of his college. James's essay, "The Moral Equivalent of War," acknowledged that

> The martial virtues, although originally gained by the race through war, are absolute and permanent human goods. Patriotic pride and

> ambition in their military form are, after all, only specifications of a more general competitive passion. They are its first form, but that is no reason for supposing them to be its last form. . . . The martial type of character can be bred without war. . . . The only thing needed henceforward is to inflame the civic temper as past history has inflamed the military temper.

James seemed to think that morality had evolved—that military forms should no longer apply themselves to armed conflict, but to improving the material conditions of life. National unity and racial improvement can be sustained without war. The problem is not the martial spirit itself, but the way it is channeled. James's notion of the moral equivalent of war did not impress Roosevelt but deeply influenced two later presidents. Inspired by James, FDR's Civilian Conservation Corps sent unemployed young men and some women into the woods to work on reforestation and soil management projects for thirty dollars a month. Jimmy Carter's allusion to the moral equivalence idea in his ill-fated national address in 1977 intended to shake Americans out of what he perceived to be a national funk, but instead caused a political backlash that boosted the fortunes of a conservative movement led by Ronald Reagan. The botched intent of the speech was to advance a national energy plan, which, if enacted, would have saved the country much grief over the next decades.

Progressives agreed that progress fostered moral improvement, but they disagreed about what counted as progressive methods. In education, the progressive course seemed clearer: as codified by an educational movement with Dewey at its head, the pragmatic theory of knowledge guided the schools of the future. In keeping with the general notion that ideas must be tested against experience and that science is only an intensified version of all the acquisition of knowledge, the progressive educators emphasized ways of learning rather than substantive content. Classrooms must be arranged so as to stimulate rather than discourage a sense of adventure in learning. Children should be educated as part of a community of discoverers. Innovations like science labs and science fairs are remnants of progressive education, but the relatively free form approach of progressive education has given way to a test-oriented culture.

The Progressive Legacy

As a popular movement, progressivism lost much of its energy during the 1930s when the post–Civil War era of economic expansion gave way to contraction. FDR was no trustbuster, and the main opposition to his early New Deal "make work" programs found most of their opposition from unions. But he found it politically convenient to implement some progressive ideas, like civic solidarity, in the midst of the national emergency. The momentum in favor of adjusting the American system to institute some public welfare arrangements advocated by progressives irrevocably changed political assumptions. In particular, by the early 1950s, all but a minority of small-government conservatives agreed that a government larger than that before World War II was acceptable in order to provide certain minimal services. FDR's "pay as you go" concept of prospective income-tax collection and the assured revenue flow it guaranteed was a stalking horse for more federal bureaucracy, including reliable and expanded funding for basic scientific research. After World War II, government-sponsored science and especially the university system benefitted from the gradual acceptance of a larger role for the state in public life.

Before the 1940s few advocated a large centralized federal bureaucracy. President Eisenhower and the East Coast establishment of the Republican Party even expanded the scope of the Social Security system. Medicare, Medicaid, and civil rights and voting rights legislation would have to wait for Lyndon Johnson, whose presidency was, in its domestic policy, one of the most progressive in history. Besides conservative intellectuals gathering under the banner of William F. Buckley's *National Review*, the last holdouts against "big government" in the 1960s were Johnson's former allies, Southern Democrats who insisted on states' rights in a last-ditch effort to resist federally imposed integration. But in spite of their race-baiting rhetoric, the Southern Democrats were not necessarily opposed to the social welfare programs of the New Deal. On the contrary, many were populists who supported government as a hedge against business interests and a source of public largesse for their impoverished constituents. The election of Ronald Reagan in 1980 crystallized the country's historic retreat from the rhetoric of progressive ideals, but not necessarily from progressive practices. American progressivism was always more oriented toward reform of capitalism than the introduction of socialism, using government to limit unjust inequalities on a case-by-case basis rather than establishing a comprehensive welfare state.

One element of progressivism that has survived the ebbs and flows of liberal ideology and criticisms of government size is its role in science. Owing in part to national emergencies that justified government investment in technology and the way that discovery fits the American frontier narrative, America's standing in science is one of the great success stories of the past century. The United States was desperately unprepared for World War II. Not only had the armed forces been greatly depleted and industry dedicated to consumer goods, research and development capacity was scattered through a few universities and industrial research institutes with little coordination or connection to the military establishment. Academic leaders resisted government grants for fear of losing control of their institutional mission; what science there was to do wasn't usually expensive enough to justify it. In retrospect, the organization of research and development was one of the Roosevelt administration's key successes that laid the groundwork for the American economy in the second half of the twentieth century. The Manhattan Project that developed the atomic bomb was the first and is still the paradigmatic "big physics" project—one to which what might be called the first "big biology" effort, the Human Genome Project of the 1990s, is often compared. But the bomb is only the most memorable technical achievement of that period, as important research and development took place in virtually every field, including medical science. White House–sponsored medical projects included new malaria therapies and studies of ionizing radiation.

Though the post–World War II period that began with virtually every other major country in ruins or bankrupt or both will not, we may hope, be repeated, current U.S. hegemony in the life sciences is still riding on the extraordinary advantages it enjoyed since 1945. The spoils of war included not only new markets but much less well-known access to enemy industrial breakthroughs like synthetic rubber and more efficient audio recording systems, without the encumbrance of intellectual property limitations. Presidential advisor Vannevar Bush made the classic impassioned and persuasive case for the federal government's continued support for fundamental research in his 1945 report to the president, *Science: The Endless Frontier*. His argument astutely incorporated familiar themes of an American narrative of innovation and progress:

> It has been basic United States policy that Government should foster the opening of new frontiers. It opened the seas to clipper ships and furnished land for pioneers. Although these frontiers have more or less disappeared, the frontier of science remains. It is in keeping with the American tradition—one which has made the United States great—that new frontiers shall be made accessible for development by all American citizens.
>
> Moreover, since health, well-being, and security are proper concerns of Government, scientific progress is, and must be, of vital interest to Government. Without scientific progress the national health would deteriorate; without scientific progress we could not hope for improvement in our standard of living or for an increased number of jobs for our citizens; and without scientific progress we could not have maintained our liberties against tyranny.

Bush's writings helped institutionalize a notion of science as a chapter in the narrative of America's national destiny, paving the way for a smart bet on technological innovation during the Cold War. In retrospect, Americans were also lucky in their opponent, for the Soviets were unable to sustain the financial resources and intellectual openness required for building a comprehensive research system that could exploit the many opportunities presented by modern science. Their successes were spectacular but narrow, expensive, and self-defeating. Hydrogen bombs, intercontinental ballistic missiles, and an initially successful space program were geared toward national security purposes, not the fundamental quest for understanding the way the world works or the tinkering that so often leads to new ways of satisfying consumer demand or creating new value. Neither Nazi Germany nor the Soviet Union could tolerate intellectual openness or establish a research and development system. Thus is fed the innovation that ultimately undergirds superpower and prosperity. Again, however, that lucky preeminence is under threat as never before, partly because science is now a global endeavor.

Vannevar Bush's most prominent legacy might be the space program. When asked to name the country's greatest achievement of the twentieth century, Americans' most popular choice is the moon landing.

Strictly speaking, getting to the moon so quickly was more an engineering triumph than a scientific one, and Bush himself later opposed the program as reckless. But thanks partly to his vision, unlike the Soviets the United States had a varied and adaptable physical and intellectual science infrastructure. The U.S. higher education system was also vastly more flexible: the top Soviet mathematicians and theoretical physicists were isolated in specialized institutes while their American colleagues held faculty positions in comprehensive and vigorous universities surrounded by energetic young students. Fundamental knowledge was of concern to all.

Less clear is whether Bush's claim that basic research should take precedence over targeted goals has been borne out as the best bet for spinning off new technologies; that aspect of his approach has lately been a matter of debate among science policy experts. The assumption that the research and development system that developed during World War II is the best path to solving social problems, Daniel Sarewitz argues, is unproven. He proposes tighter connections between research and societal goals and values. Others advocate targeted incentives for public-private partnerships, such as "tech transfer," generally regarded as a key source of America's economic growth in the late twentith century, or multimillion-dollar prizes for specific engineering accomplishments like new fuel cells. Although originally intended to encourage the development of computing technology, the life sciences have especially benefitted from the passage of the Bayh-Dole Act in 1980, which permits universities and small businesses to claim patent rights over useful inventions that were developed using federal research funds.

Although the ultimate social rewards of scientific investment may be distant or unclear, government funding of basic research is well tolerated, even popular. The National Institutes of Health is a favorite of both Congress and the public, and the National Science Foundation keeps basic physics and engineering afloat. Though our research and development investment has tended lately to drift below the benchmark three percent of gross domestic product, Americans' support of spending on science remains strong, especially if they believe there are concrete payoffs in sight. Because human beings show no sign of answering all the questions the natural world presents, there will always be more to learn and more opportunity to fund science, a fact that makes it difficult to know how much scientific investment is enough.

Whatever reservations Americans might have about the implications of science or the trustworthiness of scientists, at least in practical terms they are enthusiastic about the need to continue scientific research. According to Research!America, eighty-eight percent agree that research drives job creation and incomes, and ninety-two percent believe it is important for the U.S. economy. This has been the steady consensus in favor of science and technology investment over the past seventy years, in both the public and private sectors. As a result of that investment and continued public support, the United States will probably continue to be the preeminent center for the expansion of knowledge for the next couple of decades, barring some catastrophe that would take out much more than our laboratories. However, it seems all but inevitable that the distance between America and its nearest competitors will be vastly narrowed, putting unfamiliar pressure on this country's ability both to innovate efficiently and to cooperate with others in targeted fields. Though Americans have resolved themselves to the value of governmental investment, they no longer identify scientific and technological progress with moral progress, as was the case a century ago.

The sociologist John Evans has found that these moral reservations do not apply to chemistry and physics so much as they do to biology. The prominence of the life sciences in the twenty-first century is important for religious conservatives because of the challenges it presents to specific but basic questions like the origins of human beings. This attitude may be temporary, but in the meantime it does not make the tough choices about national investment in the life sciences any easier. A number of countries are pursuing industrial policies that aim to lay the foundations of their future economies. China, Canada, Australia, Taiwan, Singapore, and South Korea are among those that have embraced the opportunity to create new markets and new wealth based on a biotech platform. As the United States responds to this competition, the new biology will raise new questions about the place of science in the evolving American identity.

CHAPTER TWO

THE POLITICS OF HEREDITY

THE DECLINE OF FEUDALISM AND INCREASING URBANIZATION IN fifteenth-century Europe presented novel challenges to the governments of the time, such as they were. Those who could not gain entrance to the city camped out in rough shelters up against the city walls, literally the first "sub-urbs." Theirs was not a protest against exclusion so much as an implicit demand for urban growth. Gradually the walls became irrelevant, came down, or remained as curiosities for future tourists. As a result of this pilgrimage from country to town, the former intimate family ties no longer had the same influence over behavior.

On one hand, the sovereign benefitted from market towns as new sources of wealth; on the other hand, these unfamiliar concentrations of population could be unruly. And, as the catastrophic Black Death had taught, they were seedbeds of disease. The urbanization trend continued and, if anything, intensified with the ultimate decline of feudalism through the eighteeenth and nineteenth centuries, which freed up new workers to feed industry as market towns evolved into factory towns as well. With these demographic shifts in the eighteenth century, a new idea emerged that viewed populations as governed by their own logic and reason; in essence, natural collectives of human beings are more than the sum of the individuals who make them up. These larger, densely packed, dynamic, and often restive populations thus became a political problem—an object of the art of government or, in Foucault's words, a problem for *governmentality*. They were "understood in the broad sense of techniques and procedures for directing human behavior. Government of children, government of souls and consciences, government of a household, of a state, or of oneself."

The word *governmentality* is a portmanteau that incorporates the concepts of government and mentality; it refers to a certain way of thinking. It is tempting to view this new art of government as an imposition by the powerful, as libertarians might, or as some sort of social contract theory, as liberals might. But the mentality of government is not an assertion of power in a pejorative sense, nor does it presuppose a tacit theoretical social agreement. Rather, it is a society-affirming notion of control embodied in a number of seemingly beneficent institutions intended to make it possible for people to live together harmoniously and productively. The complex institutions that characterize post-Enlightenment societies—such as universities, hospitals, and prisons—are justified in terms of their abilities to realize aggregate societal goals. The same can be said of less tangible institutions like marriage and family.

As the ways that people lived together changed, as market towns blossomed and individualism took hold, learning and social intercourse intensified and capitalism matured. The traditional command approach of sovereign power had practical limitations. The emerging art of government was thus largely a post-Enlightenment internalized psychological construction that arose organically along with changed social conditions. It was integrated into a wide range of human behaviors. Before the Enlightenment, governing largely involved power over life and death. After the Enlightenment, the object of governing was *ways of living*: for example, deciding how to handle threats to public health and how to educate, determining who is sane and who not, settling what actions or conditions are offensive or dangerous and warrant incarceration, and even establishing acceptable and unacceptable forms of sexual expression.

In an immediate and obvious sense, this mentality of government satisfied the need for the sovereign to exercise power in order to maintain security. Market towns and the new systems that were their essence presented new challenges for social stability. The whole point of markets and the source of their unique energy is exchange; markets, after all, are living things. For these towns to survive, let alone to prosper and grow, there had to be physical and social movement, which comprised the circulation of individual human beings carrying goods, services, and ideas. While traditional sovereignty meant power over static spaces and populations, now both continual movement and control had to be achieved simultaneously. As a result, the idea of security itself became the idea of normalcy, or of the regularities that could, in principle, be statistically represented. In such a dynamic living system, deviance (crime, disease, mental illness) is unavoidable but must be kept within bounds of normalcy. The notion of the normal undergirds assumptions about the need for institutions of discipline, of control over these living organisms of population and the political contentions that attend this exercise of power.

Novel emerging social arrangements required a rational justification for authority, one that could be internalized by the population. Neither sovereign command nor family ties were enough. So the liberal state arose and was imbued with the new rationale for population control. Again, however, this was a not deliberate, "top down" program imposed by something like a ruling class on lower orders; on the contrary, individuals and populations integrated the mentality of government. The liberal state required a rationale, that collective, internalized notion of the legitimacy of government in the most general sense. As Nikolas Rose and his colleagues point out, "liberalism is not so much a substantive doctrine of how to govern. Rather, it is an art of governing that arises as a critique of excessive government—a search for a technology of government that can address the recurrent complaint that authorities are governing too much."

The mentality of government is therefore not a product of the liberal state but of altered social conditions. It is more accurate to say that the liberal state arose along with the need for an expanded idea of government, that it was in effect a beneficiary of this need, and it sought to exercise power in order to maintain public security. Various authorities other than the state are involved in government. Depending on a number of factors, such as the site of the activity and its objectives, each may govern according to its own logic and using its own technique.

Control over modern science and technology in particular necessarily involves other sources of governance than the state, especially professional organizations that provide expert-based standards, systems for intellectual exchange, and sanctions for misbehavior. These modes of extended governance are themselves shaped partly by the content of the activity. Thus, when physicians were left mainly to their own devices with minimal material and intellectual support for the study and practice of medicine, small guildlike forms of organization were sufficient for collegial interaction and control. The knowledge and innovation systems of the modern life sciences require social support on a vast scale, one that often engenders conflicts about its applications, goals, and practices.

Biology in Politics

"Eugenics" is one of those words that we all think we know: some form of deliberate improvement of the human race. But because it is applied to so many policies and practices spanning thousands of years and dozens of societies, it is also one of those words that can make us think its meaning is clearer than it really is. Daniel Kevles's authoritative history, *In the Name of Eugenics*, captures the many uses of the term. For all its ambiguity, the word carries a lot of baggage, especially in the era of the new biology. Its use obscures more than it clarifies. "A quick internet search identifies [eugenics] as the invective *du jour* in public discourse, shorthand for everything evil," Paul Lombardo writes. This is especially true when it is linked with both the shameful history of sterilization in the United States and the horrors of Nazi Germany. Both bioconservatives and bioprogressives worry that modern genetics is but the newest version of eugenics, but others view genetics as truly scientific and unfairly tarred with the same brush as eugenics. The underlying vagueness of the idea remains, so that often the champions and the critics of modern genetics seem to be talking past each other. Perhaps the safest generalization about the practices that are labeled eugenics is that they have something to do with a struggle for control over heredity. This has been the most historically prominent occasion for biopolitics.

In the history of philosophy, the death of Socrates exemplifies the ultimate form of pre-Enlightenment state discipline before the advent of more subtle and rationalized forms of governmentality. The right of the state to enforce discipline through its sovereign power over the lives of its citizens was embraced by the great man himself. Socrates argued that Athens had the right to take his life, however erroneous the jury's judgment about his alleged defamation of the gods. To resist Athens's authority, he reasoned in *The Crito*, would have been to deny the legitimacy of the state itself, so essential to the nature of the state is its sovereign power over the life and death of its citizens. Many classical philosophers similarly argued that the lives of citizens belong not to themselves but to the state. Without citizens, they reasoned, there would be no state, and the state thus has the right to enforce its interest in survival. In Anglo-Saxon law, antisuicide statutes may be of Christian origin but drawn from some of the ancients, like the Pythagoreans who seemed to believe in the sanctity of life. One source of the seemingly odd legal statutes that make suicide a criminal offense is that in criminalizing suicide, the state is defending its sovereignty. It is not difficult to see how this analysis may be applied to the needs of states to defend themselves from what they perceive as existential threats or merely threats to their rights or basic interests, such as the power to conscript citizens for military service and tolerate the deaths of noncombatants, including its own civilians, in the course of defense. Again, both the nature of the threat and its management require a response in terms of control over bodies and populations.

Considering the state's inherent interest in matters of life and death, the management of putatively heritable traits undertaken for the sake of the community is the paradigmatic case of governmentality. The idea that political life can and should be shaped by the application of biological principles, by biology in politics, is an ancient one. This idea is traceable once again to Plato, who advocates restrictions on marriage between those considered undesirable. In *The Republic*, Plato alludes admiringly to the far more dramatic Spartan practice of exposing weaker infants to the elements. Plato thought that the Spartans were on to something. Yet in his ideal state, physical superiority and good health was only part of the point. A natural aristocracy of those widely appreciated to have the most insight into the good, the virtue required for governance, was to be cultivated. The Socrates of *The Republic* is no romantic about reproductive matters. He speculates that a breeding lottery could accompany a fertility festival, with lots drawn to determine who among the male guardians of the state would mate with which female guardians. However, it would be important, for the actual results to be nonrandom, in order to ensure that the best mate with the best—

gold souls with gold, silver with silver, bronze with bronze. The unsatisfactory but least compromising solution seems to be a system secretly fixed by the rulers.

Plato's scenario is clearly an appreciation of the inherent potential of a politics of biology. Though Plato offered no science of the underlying reproductive mechanism, nor did he evince any interest in finding it out, his notion comprehended the common observation that some are naturally gifted. It was hard to say exactly how the somatic and the intellectual were entwined (and in fact it still is). For practical purposes, this was unnecessary to know, for the general idea that personal excellence is in some way a matter of native disposition seems undeniable. Plato was far too subtle and observant to suppose that the business of producing a great soul was only a matter of nature; indeed, in other dialogues he notes that gifted men often fail to pass on their fine qualities to their sons, though they surely would if they could. It seems part of the Platonic answer to the puzzle is the right sort of nurturing institutions. But nurture could not alone suffice. Plato's applied biology laid down a marker for all subsequent architects of Western social institutions: some thought had to be given to cultivating and enhancing certain qualities that were so important to public life. For Plato that sort of planning was plainly a state function, as the state is the res publica, the quintessential public thing.

Power over biological processes has still more insidious implications in modernity than it did in the ancient world. Ours is an era with the potential for state control over ways of living as well as of sovereign control over life and death. Today any state policies that attempt to exert control over heredity are lumped into the category of eugenics. As a result, we now tend to see the practices advocated in *The Republic* through the lens of Nazism, thereby losing the vastly different contexts. For the Nazis, race theory was both a way to explain how the German nation had been undermined by racial mixing, and ultimately to promote genocide as the strategy for addressing this "problem." Racial corruption was seen as the greatest possible threat to the purity of the *Volk*, the abstract essence of the German people, and relatedly the otherwise inexplicable collapse of the German military in World War I —the "stab in the back" blamed on international Jewish interests. Hitler declared in *Mein Kampf* that these toxic elements needed to be purged. In the Third Reich, medical aspirations and aspirations for healthy living were themselves in a poisonous mix with traditional Prussian "blood and soil" militarism, tribal mythology, and deep-seated anti-Semitism, leading to what might seem to be contradictory policies. Public health during the Nazi regime emphasized physical vigor through the consumption of "natural" foods, exercise, and resistance to tobacco and alcohol use. In parallel, cleansing the national body of foreign racial elements was a pre0requisite for the Thousand Year Reich.

The Third Reich was the reductio ad absurdum of the biopolitical state. The politics of biology plays a secondary role in *The Republic*; in the Third Reich, it has the leading part. Accordingly, doctors held a special position in promoting the health of the German body, both individually and collectively. Responding to the elevated place Hitler's ideology had given them, doctors boasted the highest proportion of National Socialist Party membership among the professions. Hitler gathered around him numerous physicians, one of whom, Karl Brandt, was a self-described idealist who was hanged as a war criminal for his part in the "euthanasia" program of the disabled and for arranging concentration camp experiments. In the death camps themselves, doctors were everywhere, giving their imprimatur to the proceedings and reassuring the defenseless victims. They needed to be present, of course, to ensure that the sanitizing of the body politic was proceeding as required and according to "professional" standards. In some cases, they also proposed and participated in the horrid experiments that were supposed to contribute to medical science, some of which were designed to address medical concerns that arose in the course of the Nazi German war effort. Nazi public health philosophy combined the traditional state control of life and death with the modern state exercise of control over ways of living.

Progress Imperfect

It is well known that in taking power, the Nazis profited from a witch's brew of political and economic factors, including a splintered and weakened political system, the postwar German depression, and a general sense of national humiliation at the hands of the British and French. Hitler also benefitted from ambivalence about science that was apparent more than one hundred years before, even as its immense advantages were on the verge of being realized. That the conditions of human life could be improved and that those improvements could continue indefinitely were not familiar concepts before the seventeenth century. Changes that could improve living conditions or expand the reach of human knowledge were not part of the elite or popular imagination. From the early to late Middle Ages, there was little reason for anyone to think in terms of improvements in this life because there had been so few. Though there were assuredly improvements in crafts like architecture, for over one thousand years the innovations that most significantly eased daily life, including the windmill and the waterwheel, could be counted on the fingers of one hand.

Then, in astonishing and somewhat inexplicable succession came the new mathematical physics, the telescope, the printing press and the information exchange it enhanced, public institutions like coffeehouses and salons, the growth of wealth through capitalism, and the rise of the nation-state. Taken together, these milestones contributed to the sense of a public sphere in which there is not only the opportunity for improvement in the conditions of human life, but a common interest in such progress. That progress turned on knowledge and demonstration rather than on metaphysical speculation. From the earliest days of the French Academy of Sciences, founded in 1666, scientists were more numerous than clergy and were described in its charter as "the most useful of all citizens."

Through the eighteenth century, the Enlightenment philosophers largely set the tone of growing admiration among the educated classes for the importance of science for social improvement. By the early nineteenth century, the growth of knowledge itself provoked anxiety. Since its publication in 1818, Mary Shelley's *Frankenstein* has been a touchstone of popular resentment of overreaching science and scientists. Lately, it has functioned as a standard reference point for the critiques of an arrogant scientific community. Although the creature might be the central figure in our nightmares (my mother remembers walking home alone in terror after seeing the just-released 1931 Hollywood version), the center of the story is the scientist, Dr. Frankenstein. The creature is a physical horror but an innocent; the scientist, with his Promethean overreaching, is a moral monster. He and his ilk are the ones we have to fear.

Shelley's tale is nearly always taken out of her context. Like the later genre called science fiction that she helped inspire, *Frankenstein* is a commentary more on the present than on the future. Shelley wrote in an atmosphere stirred by the British Romantic science movement of that period, the experimental analogue of the Romantic poets, who were excitedly investigating the properties of air, water, heat and electricity. They hoped that greater knowledge of the physical world would lead to radical social improvements such as more effective medical treatments. Finally, they believed that knowledge grounded in demonstrative experiments would liberate the human mind. The idea that electricity could reanimate a dead or assembled body was only one element of Romantic science. It is unfortunate that the most famous reflection on the Romantic scientists is a horror story. Not only did the Romantics anticipate some of the experimentation that flowered a century later with more powerful devices and more nuanced theory, they also stumbled on at least one discovery that would prove to be enormously beneficial: nitrous oxide, the first truly effective pain reliever. Thus, already in *Frankenstein* there is a template: abstract but dramatic anxiety about the direction of science that obscures its concrete achievements for improving human life. Even then scientists needed better public relations.

The appearance of Shelley's novel also coincided with the emergence of the word *science* in something like its modern meaning—as closely identified with manipulation in addition to observation in order to gain knowledge. The transformation was not immediate. In medicine, the stethoscope brought physicians into physical contact with patients. Bowel and gut sounds were traditionally thought to be a distraction from the detached observation required of the true physician. The gradual acceptance of the notion that these audible bodily emissions might be diagnostically informative urged a change in mindset that began to blur the distinction between physicians, who by tradition were to diagnose and treat from a distance, and surgeons, who engaged in invasive treatments. Experimental manipulation of largely invisible events was a likely next step, one that the Romantics foresaw. American physician William Beaumont settled the question of whether digestion was a mechanical or chemical process by passing and retrieving food into the abdomen of Alexis St. Martin, who was the victim of an accidental gunshot wound that failed to heal properly. The unfortunate St. Martin developed a stomach fistula that made him an excellent subject for Beaumont's experiments in gastric physiology.

Practitioners were aware that the more aggressive attitude of modern medical science toward the objects of investigation raised ethical issues. Beaumont himself wrote an early code of medical ethics, but he failed to take into account a conflict of interest in his financial support of St. Martin. Once it was appreciated that the ability to manipulate natural variables could create powerful insights about causation, experimentation itself came to be seen as a moral imperative. The great French physiologist Claude Bernard argued that before doctors adopt new practices, human experiments are morally required.

No single cultural figure embodies the paradoxical public attitude about scientific progress that grew out of the eighteenth century as much as Irish dramatist George Bernard Shaw, best known to Americans for his play *Pygmalion*, the basis of the musical *My Fair Lady*. Shaw championed various progressive and socialist causes, including opposition to animal experiments, then known as antivivisectionism. In 1913 he wrote of "[t]he . . . folly which sees in the child nothing more than the vivisector sees in a guinea pig: something to experiment on with a view to rearranging the world." Anglo-American culture was quick to pick up the term "human guinea pig." Shaw also opposed smallpox vaccination despite almost having died from the disease as a child. But he favored selective mating on the grounds that inferior peoples were breeding more quickly than superior ones, and he even proposed, perhaps facetiously, that people should regularly have to appear before a special committee to defend their continued existence. (Talk about death panels! This is not exactly the material for "I Could Have Danced All Night.")

The logic that enabled Shaw, other Britons, and Progressive Era Americans to approve of systematic human breeding but oppose animal experiments and vaccination was a combination of utilitarianism, an impulse for social improvement, and suspicion of biological scientists—the latter perhaps rooted in old anxieties about anatomists stealing bodies from potter's field. Experiments and vaccination require an elite individual with special expertise (and expertise is necessarily mysterious to nonexperts) to decide when they are called for and to perform them. Population improvement through breeding seemed obvious, especially if the elite advocates of such policies set themselves as the standard. As the historian Daniel Kevles observed, "[E]ugenicists identified human worth with the qualities they presumed themselves to possess—the sort that facilitated passage through schools, universities and professional training." It just happened that the talk of evolution bolstered these observations.

At the time, the scientific basis for eugenics was widely accepted, with multiple seemingly complementary sources of evidence. Some unearthed the previous pea pod experiments of Austrian monk Gregor Mendel, which demonstrated the existence of heritable traits, as evidence that human breeding could be controlled to select for the best possible genes. Then there were biostatisticians, including Karl Pearson in London, who believed that applying robust mathematical methods to better characterize evolution could also ultimately enable us to guide its direction. And there was Darwin's cousin, Sir Francis Galton, who thought that by counting distinguished Britons who were related to one another, people could arrive at an estimate of their "natural ability." Surely the prominence of these thinkers showed that they were of a superior type. Therefore, social improvement demanded the reduction of "undesirables." There were also many notable Americans eugenicist thinkers, such as Charles Davenport, founder of the Eugenics Records Office at the Cold Spring Harbor Laboratory.

Galton coined the term, but the phrase "survival of the fittest" was popularized by the philosopher Herbert Spencer. The phrase has become identified with Darwin but actually was part of Spencer's own rival evolutionary theory, based on the biological philosophy of Jean-Baptiste Lamarck, who believed that body characteristics are transmitted from one generation to the next based on their use or disuse. Spencer's theory had little to do with Darwin; for Spencer, evolution was virtually raised to the level of a metaphysical category governing all reality. According to Spencer's philosophy, everything is progressively evolving from homogeneity to heterogeneity and a higher level of both complexity and integration. Applied to social progress, this system required that nothing unnatural interfere with the evolutionary process. Thus, in an implication that would have warmed Scrooge himself, interventions that make the unfit (e.g., the poor) artificially more likely to survive should be avoided.

In addition to the general educated public, some of the most important nineteenth century thinkers without direct acquaintance with Darwin's work understood him through the eugenic lenses of Galton and Spencer. Thinkers as different as Charles Peirce and Friedrich Nietzsche opposed what they took to be "Darwinism"—Peirce because he thought it was not Christian enough, Nietzsche because he thought it was too Christian because it held out hope that humans could evolve to higher types. Somehow the phrase "social Darwinism" entered the discourse, rapidly becoming a biological philosophy believed suitable for implementation through centralized state planning. Some of America's leading early twentieth-century progressives, including Theodore Roosevelt, Woodrow Wilson, and Margaret Sanger (founder of the American Birth Control League, which would become Planned Parenthood) embraced the notion that society's burden of morally debilitated persons could be lessened through selective reproduction. Eugenics was the classic case of a blurred line between social reform and social engineering according to a set of flawed assumptions. As a popular movement it satisfied numerous agendas, as summarized by Garland Allen:

> The early 20th-century eugenics movement was a product of a particular economic, social, and scientific context: a highly transitional period in American economic and industrial expansion, the advent of a new genetic paradigm, and the ideology of rational management by scientifically trained experts. As historian Sheila Weiss has emphasized, there was enough logic to the eugenic argument—saving the hard-pressed taxpayer the burden of supporting masses of supposedly defective people—to give it popular appeal. For a segment of the biological community, it provided career opportunities that could be justified as the direct application of their science to the solution of social problems. For the wealthy benefactors that supported eugenics, such as the Carnegie, Rockefeller, Harriman, and Kellogg philanthropies, eugenics provided a means of social control in a period of unprecedented upheaval and violence.

There were modest exceptions to the enthusiasm for hereditary solutions to cure all social ills. The small first generation of well-trained American laboratory biologists had a different mission from the animal breeders and physicians: they were in search of evidence, not social control. Nonetheless, their ambivalence about eugenics is telling. In 1924, Harvard geneticist William Castle said, "We are scarcely as yet in a position to do more than make ourselves ridiculous in this matter. We are no more in a position to control eugenics than the tides of the ocean." However, Castle's textbook *Genetics and*

Eugenics, first published in 1916, urged young people who were not genetically qualified for marriage to "fulfill the racial obligation vicariously by helping to care for and to educate the children of their more fortunate fellows."

Eugenic Sterilization

In 1914, Harry Laughlin of the Eugenics Records Office at Cold Spring Harbor Laboratory published a "Model Eugenical Sterilization Law" that would authorize sterilization of what he listed as the feebleminded, insane, criminalistic, epileptic, inebriate, diseased, blind, deaf, deformed, and dependent, including "orphans, ne'er-do-wells, the homeless, tramps and paupers." A key justification for these procedures was that the individuals in question should not produce more of their type to burden the state. Another was that improvements in public health and nutrition during the nineteenth century had interfered with natural selection so that lesser types were surviving to reproduce as never before, gradually and dangerously lowering the quality of the human race. The implicit, and sometimes explicit, empirical assumption was that all of the objectionable conditions were heritable, which of course turned out not to be the case. But the damage was done: over the next few decades, tens of thousands of people were involuntarily sterilized.

The dispiriting story of America's pioneering contributions to legalized eugenics has been told by historian and law professor Paul Lombardo. A seventeen-year-old girl named Carrie Buck came to be the iconic figure in a test case for judicial validation of sterilization laws. Officials at a Virginia institution concluded that Carrie was "feebleminded" and sexually promiscuous, like her mother. To avoid more "socially inadequate offspring," Carrie could be sterilized under the Virginia statute. In 1927, writing for the majority in *Buck v. Bell*, Supreme Court justice and noted progressive thinker Oliver Wendell Holmes, Jr., wrote a decision that upheld the state law. "It is better for all the world," Holmes wrote, "if instead of waiting to execute degenerate offspring for crime, or to let them starve for their imbecility, society can prevent those who are manifestly unfit from continuing their kind. . . . Three generations of imbeciles are enough." A few years later, the newly installed Nazi government partly based its sterilization law on the Laughlin model, though the Germans implemented the philosophy with a vigor and commitment that vastly outdistanced American eugenics policy. Borrowing from Laughlin's "Model Law," the Nazi regime sterilized more than 350,000 people, outdoing even the tens of thousands sterilized in the United States, and recognized Laughlin with an honorary degree from the University of Heidelberg in 1936.

Many factors contributed to the attraction of early twentieth-century progressives to eugenics. As Garland Allen writes:

> Eugenics fit perfectly with Progressive ideology. Eugenicists were scientifically trained experts who sought to apply rational principles to solving the problems of antisocial and problematic behavior by seeking out the cause, in this case poor heredity. The best schooling and social training—like the best soil—was of no avail if hereditary constitution was defective. Eugenicists were to be the "managers" of the human germ plasm, in the progressive spirit, and would take control of human evolution.

Americans were obsessed with the idea of racial differences and improvement by the late nineteenth century, before the obsession was finally interrupted in the 1940s as the catastrophic consequences of racial theory and the Holocaust were realized. The national success of Anglo-Saxons in Britain and the United States, as compared to every other people and region, seemed obvious. What could be the explanation but racial differences? Some sought still greater racial perfection. Eugenics provided one path, strenuous living provided another, and military conflict, especially, offered another (unless, as some have argued, the best young male stock was lost in combat). Thus, as acting Navy secretary, Theodore Roosevelt embraced the dubious theory that the Spaniards had blown up the USS *Maine* in Havana harbor, exploiting the resulting war fever in America. As president, he pursued a guerilla war in the U.S.-controlled Philippines that left thousands of American soldiers dead. Roosevelt's expansionist ideology was fueled partly by his worry that the end of the continental frontier would impair the American character. Setting a pattern for later American presidents, he sought new frontiers to conquer. In one sense, the new worlds were inconveniently populated by people of color; in another, they provided an opportunity for strengthening the race through struggle. Looking back through the viewpoint of a century of genocide, it is difficult for us to understand the complexities of earlier notions of race and the ways they could coexist with ideas of progress. The contradictions are intriguing. Socialist progressives like H. G. Wells approved in his prophetic novels of euthanizing "the weak and sensual," while the right-wing progressive Roosevelt provoked outrage from Southerners not long after he took office for inviting Booker T. Washington to dine with him in the White House.

Behind the enthusiasm for eugenics lay an impulse to improve social conditions in the wake of an era of industrialization that brutalized and exploited many. At least two lessons stand out as parties to the new politics of biology. One is Charles Peirce's idea of fallibilism: with the possible exceptions of logic and mathematics, no scientific account of the world of experience can achieve absolute precision. A second lesson of the eugenics movement is that biopolitical actors, especially states that possess police power, must avoid identifying with and enforcing a particular biological philosophy that could infringe on the rights of some members of society. Taken together, the possible results of policies unconstrained by fallibilism and excessive state identification with a certain biological philosophy were vividly illustrated by state-sanctioned sterilization in America and "scientifically" justified genocide in Germany. So, too, were they shown through the fraudulent claims initiated in the midst of the 1920s Soviet famine by the agronomist Trofim Lysenko. As Lysenko persuaded Communist authorities that acquired characteristics could be inherited, the result was the censure and deaths of hundreds of opponents, the cessation of scientific research on genetics, and the ultimate setback of Russian biology for generations. In a bizarre twist, Lysenko did not apply his anti-Mendelian theory to human inheritance and even criticized eugenics as "bourgeois pseudoscience," perhaps because the Nazis had so vigorously adopted a eugenic interpretation of genetics.

After World War II, the word *eugenics* acquired its current bad odor. Modern geneticists are loath to accept any association with the movement. Yet the fact remains that eugenics was considered legitimate science by influential academics and intellectuals irrespective of their other political views. Although usually identified with the Progressive Era, a right-wing eugenics movement persisted long after the progressive movement faded among radical conservatives who detested FDR and the New Deal. Distressed at the dark shadow that had been cast over their movement, they longed for the day that some semblance of the undertaking would be restored. And there are still those who stubbornly hold a racial theory in search of eugenic facts. Perhaps the most important source of support for eugenics research for over seventy years has been the Pioneer Fund, identified by the Southern Poverty Law Center as a hate group.

Liberal Eugenics?

Some draw parallels between the enthusiasm among elites for eugenics a century ago and the contemporary excitement about the promise of applied biology. Others point out that there are crucial differences between our understanding of genetics before and after the decoding of the genome. Even so, these differences may occasion more alarm than comfort. While basic research has led to a vastly improved understanding of underlying mechanisms and greater diagnostic capabilities, it has also equipped scientists with the tools to screen embryos and manipulate genetic endowment in certain instances. There is the prospect of much greater control ahead. Defenders of modern genetics note that the question now is not one of state control but of personal choice, or the advent of consumer rather than state eugenics. Critics argue that, in terms of applied genetics' implications for dignity and justice, this is a distinction without a difference. These skeptics ask, if ensuring the predominance of the "best types" is not the goal, then what is?

Inflaming the critics of unrestrained biotechnology, some writers have advocated what is often called a new "liberal eugenics" that condemns "negative" approaches like sterilization and aims to keep the state out of reproductive policy. To avoid associations with political control of reproduction through negative or positive eugenics, geneticist Lee Silver coined the term *reprogenetics*. Viewing some form of parental consumer choice as inevitable given the new genetic technologies already becoming available, liberal reprogeneticists believe that it is possible to preserve reproductive freedom while also minimizing the numbers of children with congenital disorders. Some would include enhancement technologies among those techniques that parents or potential parents should be permitted to introduce, if and when they become safe and practical. Bioprogressives of various stripes, from liberals to libertarians to transhumanists, have expressed support for reprogenetics, while bioconservatives on the left and the right have for various reasons shown deep concern.

In response to reprogenetics, some conservative commentators such as Yuval Levin call attention to the left's association with science.

> American progressives have stumbled on this path before. In the late nineteenth and early twentieth centuries, the cause of material progress and scientific control, together with some crude misapplications of Darwinism, combined to form an energetic and progressive program of eugenics, beginning with public education toward selective breeding based on valued family traits, and culminating in a massive project of sterilization—including coercive sterilization laws in more than twenty states—of those found mentally or physically wanting.

A somewhat different critique of liberal eugenics comes from within the left itself. The criticism is expressed not so much as a drift to another overt catastrophe, but through the perspective of consumerist narcissism and perhaps unfair, class-based advantages for the children of the well-to-do. The communitarian Michael Sandel has suggested that the new, market-driven liberal eugenics is less dangerous but also less idealistic than the old, state-directed eugenics.

> For all its folly and darkness, the eugenics movement of the twentieth century was born of the aspiration to improve humankind, or to promote the collective welfare of entire societies. Liberal eugenics . . . is not a movement of social reform but rather a way for privileged parents to have the kind of children they want and to arm them for success in competitive society.

Some defenders of biotechnology reject the justice argument on the grounds that economies of scale and the incorporation of reprogenetics services into insurance programs will someday make technologies like preimplantation genetic diagnosis available to all. Just as sanitation and vaccination have helped level the playing field of opportunities for the less advantaged, the ability to assess embryos for genetic disorders or even to provide enhancing interventions will help equalize the potential of the next generation regardless of socioeconomic status. On this view, a rising tide of reproductive genetics lifts all boats.

Reprogeneticists see themselves as advocates of market and reproductive freedom. They also cite the fact that, class and economic privileges notwithstanding, parents are permitted to finance "improvements" in their children's competitive position with SAT tutors and tennis lessons. Suppose, for the sake of argument, that state biopower is removed from the scenario. Then aren't the critics of liberal eugenics "mystifying" biology? Aren't they assuming that biological modifications, including those engineered into embryos or selected from among embryos, are inherently different from the advantages parents are allowed to provide for their born children? Three philosophers who favor some forms of enhancement put the point this way:

> The leeway parents are generally allowed to pursue the best for their children may seem unproblematic because there is a tendency to think of their efforts as "environmental" factors that help to develop the capacities or capabilities their children already have or are capable of having. The parents are only "bringing out the best in them," or developing "the potential" that is already there. In contrast, the use of genetic information and intervention (whether somatic or germline) suggests parents are changing their children in some fundamental way, making them different from what they otherwise would have or could have been.

The term "designer babies" is an implicit criticism of genetic modification because it suggests a child is to be treated as simply a consumer item. But why assume that genetics is somehow off-limits to parental choice? Don't all parents want to "design" their babies insofar as nature will allow? Isn't that what we mean by parental nurture? Bioprogressives reject the notion that genetic interventions by parents are per se any different from other attempts by parents to give their children a better future. There could be problems of safety and fairness with all sorts of child-rearing practices, not only those tied to genetics. According to standard liberal theory, all are and ought to be subject to governmentality, especially to the extent that they create unfair inequalities.

Muscle-bound Biology?

These philosophical arguments about the power of reproductive genetics are somewhat disconnected from the complexities of the science and the prospects for its application. Embryo screening in fertility clinics is still a rare event compared to the number of live births. As it stands, screening can only be done if a mother underwent in vitro fertilization (IVF), a labor-intensive, expensive, and invasive procedure even in wealthy countries. Although numbers are hard to come by, the controversy preimplantation genetics has generated has surely exceeded the actual frequency of its use. Normally, an embryo's DNA is analyzed only in response to some concern triggered by the parents' family medical history or known genetic heritage. A prominent example involves diseases linked to the X chromosome, like Huntington's, that are relatively easy for scientists to detect in genes.

But a family of technologies now limited to the most advanced research laboratories, grouped under the heading of "microarray analysis," may someday be available to many people interested in having babies. Then prospective parents will learn about thousands of traits in each of their embryos. In one sense, this suggests a great deal of parental power over the ultimate character of their potential offspring. However, this might turn out to be a case of muscle-bound science, strong but incapable. There may be a lot of information but not a lot of knowledge.

To see how this could be so, take one product of microarray analysis called copy number variation. With computer science combining with molecular biology, so much information is being produced that increasingly advanced software is needed to make sense of it. One target of this computer-assisted biology is the number of duplicated copies of a gene, or the number of deletions of a gene, on the DNA chromosome. Researchers can compare these copies or deletions across a population. In theory, it should be possible to see what sequences are significant for a variety of diseases in a population. Already this concept has been used to identify sequences that seem to be associated with cancers, autism, schizophrenia, and Crohn's disease. If these results were reliable, one could imagine them being applied in fertility clinics to screen embryos to avoid having a child that would be at risk for these diseases. Moreover, it is not only diseases that could be identified; copy number variation research has also been applied to body mass index, for example. The reach of genetics is long, even though it is often only partial.

As good an excuse for more philosophical debate as copy number variation might be, for many reasons the technology is nowhere near ready for prime time in breeding healthier or superior beings. The analytic systems now being used were not designed for clinical use, so the results are difficult to interpret, and it is not clear which gene repeats or deletions have to do with any particular condition and which have no outward effect on traits. Only high-end research labs are using the technology, trying to sort out what their results mean. In the words of Nancy Spinner at Children's Hospital of Pennsylvania, "The problems that are now facing us are just at the very beginning of a) understanding the variation in the genome, and b) understanding how it relates to health and disease. It is all so new that it is very difficult. The science is just not there yet, to be able to tell us what it means to have something that is extra or missing." Enormous amounts of data are being generated by research studies, but what remains unknown is their relevance to the likelihood that someone who develops from an embryo might have a disorder or some specific trait months, years, or decades later.

Gradually, there will be greater understanding of the relationship between copy number variation and diseases, physical traits, and perhaps even psychological characteristics. The statistical correlations will likely be known long before the underlying mechanisms. In many cases, perfect prediction will be impossible because of the interaction between genetics and the environment that causes genes to be turned on and off. There is no way of foreseeing all the factors an individual might be exposed to over a lifetime, such as radiation and chemicals, much less how that exposure will occur, including issues of how much, when, and to what body parts. But if the stakes are considered high enough, even modest powers of prediction may generate conflict. Look for a series of struggles for access and control over these data between many interested parties: regulators, industry, researchers, advocacy organizations, and, of course, nervous prospective parents.

Personalized Genetics

While laboratory geneticists wrestle with the vagaries of data generated from the new computational biology, a different approach is illustrated by the success of direct-to-consumer genetics companies. For a fee ranging from several hundred to over a thousand dollars, these companies mail a test kit to your home for collection of a DNA sample, usually by means of a cheek swab. The consumer then mails the sample back to the lab and is informed of the results by mail, over the phone, or online. Different companies test for different factors, mainly a person's ancestral origins and the risk of getting a certain disease. Though counseling about the meaning of the results accompanies the information, ethical objections to this way of obtaining personal genetic data have been voiced, and regulators are keeping a wary eye on the several dozen companies in the personalized genomics industry.

One of the largest of these companies, 23andMe, published a study based on the genetic information of nine thousand participants who also filled out surveys about their health and physical characteristics. Their Web-based research program, called 23andWe, then looked for genes that correlate with twenty-two traits, including curly hair, freckles, photic sneeze reflex (the tendency to sneeze when entering bright light), and asparagus anosmia (the inability to smell asparagus metabolites in urine). With its access to so many individuals, the company is able to conduct these genome-wide association studies with great efficiency at low cost, challenging the traditional way of doing science that typically relies on several stages of lab studies and statistical analysis of a few dozen or perhaps a few hundred samples, with results finally published in a peer-reviewed science journal. In general people go into these studies having been told the sources of the samples go into the study told that the samples will not be linked to their names, that they will not be told the results, and that the implications of results will not be known to a sufficient degree of scientific confidence. This is the way that lab geneticists at major research centers are working on the implications of phenomena like copy number variation. The contrasting approach taken by ventures like 23andWe crunches huge amounts of data from perhaps ten times as many research subjects and presents the correlations between genes and traits in a press release and perhaps in a science journal. Many scientists object that too much irrelevant "noise" might confuse these genetic associations and that an initial hypothesis is needed to avoid a lot of misleading, incidental correlations. The traditional process emphasizes the search for underlying mechanisms that connect genes to a condition; the consumer approach emphasizes patterns, with an understanding that mechanisms may be identified much later, if ever.

What is not debatable is that genetic associations can have enormous emotional significance. They offer people some sense of control over their destinies. Google founder co-Sergey Brin, whose wife started 23andMe, offers himself as a case in point. He has a genetic alteration that puts him at higher risk for Parkinson's disease, and some studies have suggested that exercise and coffee can lower the risk. "I now have the opportunity to adjust my life to reduce those odds," Brin says. "I also have the opportunity to perform and support research into this disease long before it may affect me. And, regardless of my own health, it can help my family members as well as others." Francis Collins, director of the National Institutes of Health (NIH), who oversaw the sequencing of the complete human genome, is a prominent case in point. When he learned of his own genetic risk for diabetes, he changed his diet, began an exercise program, and lost twenty-five pounds.

By contrast, the person whose DNA was used (at first anonymously) as the basis for the complete "mapping" of the human genome, and whose private company raced with the NIH to complete the map, does not place the same stock in genetic testing results. Though he has a gene associated with a greater risk of Alzheimer's disease, scientist and entrepreneur Craig Venter says: "[I]t impresses me little because I could have dozens of other genes that counteract it. Because we do not know that, this information is meaningless . . . we have, in truth, learned nothing from the genome other than probabilities. How does a one or three percent increased risk for something translate into the clinic? It is useless information." The information might indeed be useless for medical care, except to people having trouble achieving a successful pregnancy. When they have a choice of embryos in a fertility clinic, they may well want to avoid an elevated risk of Alzheimer's.

When the journalist Ronald Bailey had his genome scanned by a couple of companies in 2010, he discovered that his lifetime risk of Alzheimer's is average, his risk of male pattern baldness low, and his risk of diabetes higher than average. As for ethnicity, Bailey's DNA analysis gave the lie to a family legend of descent from a Cherokee princess and confirmed the more prosaic truth that he is thoroughly European, with a Y chromosome that points to Ireland. As a libertarian bioprogressive, Bailey opposes governmental attempts to control consumer genetics and dismisses fears of prejudicial genetics judgments by insurance companies or employers. Rather, he believes that a more realistic concern is that state authorities will find ways to obtain and use our genes against us. With proposals like establishing a national DNA database to help protect against terrorism floating around, with DNA data perhaps ensconced on government-issued ID cards, personalized genomics is at the very least ripe for future biopolitical contention.

In one sense, the approaches behind copy number variation and personalized genomics are complementary. Until scientists unearth mechanisms and develop direct interventions, companies like 23andMe can provide access to information that people could use in changing their lifestyles. One can well imagine a scenario like this: a couple's baby was part of a major research hospital's study that revealed all sorts of genetic additions and deletions, none of which had medical significance that was well understood; then, the couple sent a sample to a company and learned of many correlations the scientists could have provided but did not think were significant. Whether the marginal benefits of this kind of information are worthwhile is a different question. After all, since all of us will get sick with something, the precise risk reduction for people like Brin and Collins who want to respond to the news of a potential impending disease, in contrast to those like Venter who do not, is unknown.

Consumers of genetic information who are sufficiently worried about, say, depression, may make reproductive decisions based on a very remote chance that they or their children might be at risk. Some diseases weigh more heavily on people's minds than others, for highly personal reasons. Although the psychology of risk is highly variable and not always reasonable, depression does seem to run in families, and anyone who has a close relative with a major depressive disorder may be forgiven for at least having second thoughts about bringing a child into the world with a chance of having this disease. The National Institute of Mental Health has identified several genes that predispose people to a higher risk, and several companies offer direct-to-consumer testing for the ones associated with depression. One can hardly blame prospective parents for wanting to know their results.

Beyond diseases, the personalized genomics world also offers insights into certain heritable advantages. One company advertises its Inborn Talent Genetic Test, which "reveals the inherited and endowed inborn talents of a child scientifically from the genetic makeup of his/her DNA. The test result will therefore help parents identify their children's hidden talents that may not be obvious at young age." Among the inborn talents tested for are IQ, EQ, and athletic ability and artistic creativity, in particular the "dancing gene" and the "musical gene." Such tests would seem useful in Nietzsche's *Thus Spoke Zarathustra*, in which the prophet announces the coming of the overman who shall be skilled in "singing, dancing and laughing."

It might be thought that common ground on the ethics of genetics can be established based on the idea of fighting disease but disallowing enhancement; in other words, therapy is acceptable, but improvement is not. But it should be clear by now that the new genetics at least challenges the distinction between therapy and enhancement: for instance, the net result of correcting for a disabling trait like depression might be an individual with exceptional resilience. There are also some applications for these categories do not even make sense. For example, a lot of lab work, and one of the reasons people seek to have their own genomes analyzed, has to do with the ways that genetics might tell us about race. So far the results are equivocal, since it is very clear that our standard racial categories are not perfectly mirrored in genetics. As explained by a group of Stanford University geneticists and ethicists, "there is no scientific basis for any claim that the pattern of human genetic variation supports hierarchically organized categories of race or ethnicity." It turns out that when groups are classed according to language, geography, or culture, there is a lot more genetic variation *within* a group than there is *between* groups.

Nonetheless, we can be sure that there will always be attempts to use genetics to support racial theory. As the Stanford group acknowledges, "a broad range of associations between genetic markers and human traits—including diseases—is emerging." Although the group points out that the associations are "only statistical," that will not prevent their use, including by companies offering personalized genome testing. The computational biology that makes it possible to search for more and more correlations between genes and traits will create more opportunities for misleading generalizations about race. Much symbolic authority will continue to reside in the combined ideas of race and genetics.

While there is no last word on the vexed and vexing word "eugenics," there can be clearer uses of it. Eugenics is the most ancient and most prominent biopolitical conflict zone, but as one takes stock of the history, one has to wonder what exactly the battle has been about. The mysteries of heredity? Social engineering? Public health? Reproductive choice? Race theory? Applied genetics? Mate selection? Enhancement? Abortion? Euthanasia? The idea of eugenics achieves its power through a combination of primitive tribalism and the longing for rational control. It has become a mirror onto which societies project their fondest aspirations and darkest anxieties. We may never liberate ourselves from its grasp, but at least we should insist that whoever uses the eugenic mirror links it to critical reflection on specific policies or practices and not as a distorted reflection of our hopes and fears.

CHAPTER THREE

DANGEROUS IDEAS

On August 9, 2001, the forty-third President of the United States delivered what was considered the most important address of his young presidency. From his ranch in Crawford, Texas, President George W. Bush announced to a national television audience his policy on a question that, along with the murder of a beautiful congressional aide and a series of shark attacks in the southeast, had fixed much of the country's attention that summer: what to do about federal funding of human embryonic stem cell research.

Although Americans had long been aware that the life sciences raise challenging ethical issues, this controversy was marked by a unique intensity. The new biopolitics had arrived. What distinguished the stem cell issue, of course, is its connection to human embryos. Nonetheless, there was something odd about so much political drama being stimulated by an arcane laboratory procedure practiced in a scattering of research sites. Compared to the decisions that led to vastly more consequential presidential decisions, like the development and finally the use of the atomic bomb, this one was not only publicly heralded, but followed a series of Oval Office meetings with some of the country's top thinkers, mainly philosophically oriented experts in bioethics. According to some accounts, these sessions took place in the presence of Karl Rove, the architect of President Bush's political career.

The New Biology

The most cited reason for the emergence of biopolitics is the media-driven, overheated, and often uncivil cultural politics of America in the late twentieth century. What has not been adequately recognized is that changes in the nature of biology and its related disciplines have helped to create the conditions for the new biopolitics. In his "Theses on Feuerbach," Karl Marx proclaimed that "Philosophers have hitherto only interpreted the world in various ways; the point is to change it." Until recently, much the same might be said of biology. For most of its history, biology (or, more comprehensively, the life sciences) has been a matter of interpretation, not change. Aristotle's *History of Animals* is the first great descriptive work in what was for millennia called natural philosophy. It begins, "Some parts of animals are simple, and these can be divided into like parts, as flesh into pieces of flesh; others are compound, and cannot be divided into hands, nor the face into faces." Scanning this work of astonishing genius, one is sympathetic to the sage observation that Aristotle was perhaps the last person to know everything there was to know in his time. It took centuries to push biology forward, but there were other geniuses. During the flowering of the caliphate in the ninth century, Al-Jahiz wrote the *Book of Animals*, describing food chains and seeming to anticipate natural selection. It took another millennium for Anton van Leeuwenhoek to improve on existing microscopes and describe spermatozoa and other tiny life forms. The eighteenth century also saw an explosion of interest in classifying living things, as in Linnaeus's taxonomy.

In a sense, when James Watson and Francis Crick decoded the DNA chromosome in 1953, they were extending this descriptive tradition. Their model, however, was based on experimentation and accomplished through new laboratory tools. The inextricable combination of theory and technique that spawned molecular genetics has already begun to make possible modifications in biological nature that were barely imagined before. Since the 1970s, recombining portions of the DNA molecule through chemical splicing has advanced to creating and synthesizing chains of genes, resulting in independent and self-replicating cells with properties that would not exist without human technologies. Biologists have come a long way from the observation and classification of flora and fauna à la Aristotle, and their tools are vastly more precise and powerful than traditional breeding of plants or animals. They are now in a position to bring wholly new creatures and even new species into being; they are not just interpreting the world, but developing the knowledge and abilities to change it.

Even if the biologists don't literally create a new world in the wake of synthetic biology, genetics, and a raft of related disciplines and technologies, the old one will never be the same. Just as the twentieth century was the age of physics, the twenty-first is the age of biology. As described by a noted scientist in 2004:

> [T]hree broad research programs will account for much of this new biological knowledge. The first is genomics. It has already reshaped biological inquiry in fundamental ways, and as genomics and proteomics meld further with developmental/cell biology and the agricultural sciences, our world will be transformed. . . . The second research area is . . . "evolutionary biodiversity science." This name describes the product of uniting evolutionary and systematic biology with the environmental sciences—namely, comparative, functional, and integrative biology whose center of gravity is understanding Earth's biological diversity at all levels of organization. . . . The third research program, bioinformatics, is the glue that will bind the other research efforts together. It will create entirely new fields of research as it integrates genomics-related sciences and evolutionary biodiversity science into earth sciences, chemistry, and other disciplines.

New organisms will be created that will treat disease and provide new sources of food and energy. We will acquire a deeper understanding of life on Earth in its marvelous complexity. Information systems will help unify knowledge across arbitrary disciplinary boundaries while creating new avenues and methods of exploration.

Once acquired, the knowledge of genes, proteins, the systems to which they belong, and the information they encode can be neither forgotten nor ignored. But what, really, can we say has fundamentally changed? After all, for decades the applications of the life sciences have been part of political discourse, as in eugenics, "test tube babies," human experiments, prenatal testing, and the like, not to mention the abortion debate with its uniquely American intensity. The fundamental difference is that up until the 1950s, biological experiments, including those involving heredity, were essentially empirical. They were "biotechnology" in the gross sense, not bioscience. After Watson and Crick published their 1953 paper in which they explained the structure of DNA, scientifically based genetic control became possible for the first time, affording the opportunity to engage in discovery through experiments that manipulated the underlying mechanisms of life.

Challenging the "Ladder of Nature"

Bioscience, the new experimental biology, challenges our most entrenched categories and certain arbitrary boundaries that help us order our experiences. Darwin showed that patient, unprejudiced observation and powerful reasoning could challenge our assumptions; at a simpler level, we may recall the shock that accompanied the seventeenth-century discovery of black swans in Australia. But there is no arguing that the pace of discovery is vastly quickened by the manipulations of variables in the laboratory. And one characteristic of the new biological knowledge that is especially disconcerting and even threatening is that it defeats our assumptions and prejudices in a special way, by blurring what we thought were clear lines in nature.

Paradoxically, our tendency to categorize and to seek clear boundaries is itself a product of evolution. One example is the line between self and not-self that babies need to learn, which Freud claimed is required in order to contain (though it never fully overcomes) our inborn narcissism. Other important boundaries are social, those that tell us who is in the group and who is not. Thus, clans and kinship provide a source of fellow-feeling and mutual support, but they also entail rivalries that can be expressed in irrational fear of the "other" and, in the extreme, race hatred. The New York University neuroscientist Elizabeth Phelps has reported on neuroimaging studies indicating that our brains respond to images of members of other racial groups differently than to those of our own. Perhaps that in itself isn't surprising; Phelps warns that we shouldn't read too much into results like this, as laboratory data cannot predict human behavior. Yet the fact that these responses appear to show up so clearly in the activation of certain brain organs like the amygdala suggests that evolution finds and reinforces convenient perceptual categories. No wonder resistance to evolutionary theories of descent, like the proposition that *Homo sapiens* shares a common ancestor with other primates, is so intense. It is part of a panoply of a priori constructed boundaries that have been, for eons, largely functional.

We love to put things in boxes. Popular science writers have even categorized infant brains as inherently male or female, despite neuroscientists' view that this notion is spurious. The fact is that categorizing is fun and as natural and comforting as can be. Although infant brains are not different depending on whether they are male or female, what we do see early on in infants is the drawing of conceptual lines. As the neuroscientist and philosopher Jay Schulkin notes, "we are taxonomic animals from our earliest infancy; we categorize objects. . . . Categories of understanding converge at every stage in our intellectual development." According to Schulkin, we are especially adept at categorizing life forms. "An orientation to biological objects seems to be a core disposition in our cephalic organization." No wonder modern biology's tendency to break down boundaries is disconcerting. The very idea of evolution seems to be counter-evolutionary.

The human desire for order is profound. Formerly philosophers and theologians spoke of a "natural ladder" or, in the language of intellectual historian Arthur Lovejoy, a "great chain of being." This chain was conceived as a divinely created hierarchical spire in which all entities had a unique place, as in steps on a ladder. At the top is God, then the angels, then man, the other creatures, and finally earth. The chain is a Platonic passage from the most perfected, pure spirit, down to the least perfect, mere matter. Even earthly matter can be viewed as a hierarchy from precious metals at the top to dirt at the bottom. Thus is order given to the universe in which everything has a place, fulfilling the Greek vision of a cosmos.

What is unclear in this ontology is whether man can add links to the chain or distorts the natural order in doing so. Do physics and chemistry upset the comforting cosmological scheme by introducing new elements to the periodic table, for example, or are these accomplishments only applications of the divine spark of intelligence in man? In creating new life forms, modern biology provides still more disconcerting counterexamples to entities once thought to have been excluded from creation. In fact, if the "chain of being" metaphor still has any resonance at all, it lies in one of the most satisfying aesthetic ironies in the history of ideas: the double helix of the DNA molecule. Less than fifty years after the decoding of the structure of DNA, the controversy about a rather mechanical set of laboratory techniques often referred to as "cloning" and human embryonic stem cell research was a transformative episode that brought the control of biology to the political foreground on an international scale.

Enter the Bioethicists

The 1953 Watson and Crick paper included a line that was perhaps the greatest example of understatement in the history of science: "It has not escaped our notice that the specific pairing we have postulated immediately suggests a possible copying mechanism for the genetic material."

It took a while for philosophers and theologians to digest the implications of that modest sentence. What it meant was clear to biologists from the start, for in the laboratory DNA by itself falls to pieces. But if one can copy and thereby amplify DNA many times over, and if one has a way of reading the genetic code, then that information can be used as the basis for remarkable experiments and discoveries. In the years after Watson and Crick's paper, biologists painstakingly developed techniques to do precisely that, creating new chemicals and devices to manufacture large amounts of DNA and methods of transferring it into other genes. All this took about two decades, but by the late 1960s it was clear to those who took the time to look that this work was rapidly advancing. The early bioethicists were among those few who took notice; their debates anticipated the new biopolitics of our own time by a generation.

One of the first of these thinkers was Joseph Fletcher, formerly a Christian ethicist and Episcopal priest. In 1974, he published *The Ethics of Genetic Control: Ending Reproductive Roulette*, a work that was and remains in many ways quite radical. Fletcher's book was written partly in response to *Fabricated Man: The Ethics of Genetic Control,* by Princeton University theologian Paul Ramsey. Fletcher even used Ramsey's subtitle as his title. Like Fletcher, Ramsey had a background as a Christian ethicist. But there the intellectual similarities ended. Ramsey saw the prospect of genetic manipulation as one more potential case of the dehumanization of modern medicine, of the tendency to turn physicians into body mechanics. Genetic power could take the objectification of human beings to an extreme: if human beings are given the power to modify others as objects, they may come to see those they manipulate as mere objects. "Man as a manipulator is too much of a God; as object, too much of a machine." For Ramsey, who taught and inspired a generation of neoconservative bioethicists, any artificial intervention in the process of procreation violates that part of man that is essentially human.

For Fletcher, on the other hand, to be civilized means precisely to be artificial. Fletcher held that, rather than become tyrannized by "reproductive roulette," we should control our biology: "[J]ust as the technology of nuclear power is on an irreversible, world-shaking course, so the discovery of behavior controls and molecular and cellular biology's grasp of how to design or alter the genetic constitution of human beings might well not only be man-shaking but irreversible, at least in some of their uses. . . . Discoveries like genetic coding make the old-fashioned hard technologies seem like child's play." We cannot walk away from the road ahead, nor, according to Fletcher, should we fear it. "It is precisely because men are sapient that they can control their biology. . . . Beyond any precedent we are now in a position to change not only the social and environmental conditions of mankind but even man himself, his very stuff."

At first glance, Fletcher might seem an unlikely advocate of deliberate control over reproduction. He had been a member of the clergy (though he became an atheist later in life), a labor organizer (he was beaten unconscious while organizing tenant farmers), and a Communist sympathizer (he was called the "red churchman" by Senator Joseph McCarthy). In the 1950s and 60s, Fletcher developed the theory of situation ethics but the term is often used dismissively as a kind of relativism that was not at all what he had in mind. In his 1966 book *Situation Ethics*, Fletcher argued that love is the only intrinsically good thing there is, therefore any action or policy must be judged in terms of its consequences for increasing love. Because he believed that too many children are unwanted, he worked with Margaret Sanger and Planned Parenthood; because he believed that unnecessary suffering at the end of life was incompatible with love, he was an active member of the Society for the Right to Die.

However one judges it, Fletcher's thoroughgoing philosophy was decades ahead of today's bioprogressives and transhumanists. Similarly, Ramsey's stern traditionalism has inspired contemporary bioconservatives on the right. In fact, the elite Fletcher-Ramsey debate of the 1970s was the intellectual precursor to the ideas underlying much of the new biopolitics as it is unfolding forty years later. Fletcher and Ramsey were of that generation that grew to maturity through world wars, the Holocaust, and eugenic delusions; the decoding of the genome opened up possibilities for the latter in vastly more technically sophisticated forms.

Hans Jonas, philosopher at the New School for Social Research in New York City, was another member of that generation whose youth was even more directly shaped by those midcentury events. Also a larger-than-life figure, Jonas was a German Jew who studied under Martin Heidegger and was a close friend of fellow student Hannah Arendt, a Jew who became Heidegger's lover and defender against later charges of Nazi sympathies. With the rise of Nazism, Jonas left Germany for Palestine and fought with the British Army in Italy. After the war, Jonas discovered that his mother had been killed in the Holocaust. He returned to Palestine to fight for Israeli independence. Years later, he emigrated to the United States.

Jonas's approach to the era of genetic technology was somewhat different from those of Fletcher and Ramsey. In *The Imperative of Responsibility*, Jonas emphasized the importance of recognizing our collective limits in predicting the implications of new scientific knowledge, which Jonas called "predictive wisdom." Practical control over the human genome is an event that especially calls upon the virtue of humility, for that genome is the product of eons of trial and error that we cannot hope to reproduce or fully understand. Jonas also formulated a view about genetic modification that has influenced thinkers on the right and the left. In essence, he worried that control over another's genome could unfairly limit their future options by modifying their abilities. One version of this argument is that we are all entitled to an "open future," in which we deserve to have the opportunity to exercise our choices in life depending both on circumstances and on our inherent potential. But is such genetic control possible? Certainly there were geneticists in the 1980s who believed that the genome did hold the key to such mastery, and this was one of the notions that helped promote the Human Genome Project. Now, however, it is clear that the kind of targeted control that worried Jonas and others is not available through knowledge of the genome alone. Perhaps, given time and resources enough, the proteins that genes construct to do the grunt work of sustaining living things will be manageable in this way. That is the new frontier in genetics, but even a modicum of that route of controlling human characteristics is very far away.

The echoes of these academic discussions in the 1970s and 1980s have in the intervening years become far more angry and shrill. For example, in 1997 a group of distinguished scientists and intellectuals, including Francis Crick, Isaiah Berlin, and Kurt Vonnegut, defended the prospect of human cloning in a way that seemed to attack traditional values implicitly through the challenge to natural boundaries represented by experimental biology.

> What moral issues would human cloning raise? Some world religions teach that human beings are fundamentally different from other mammals . . . . Human nature is held to be unique and sacred. Scientific advances which pose a perceived risk of altering this "nature" are angrily opposed. . . . [But] [a]s far as the scientific enterprise can determine . . . [h]uman capabilities appear to differ in degree, not in kind, from those found among the higher animals. Humanity's rich repertoire of thoughts, feelings, aspirations, and hopes seems to arise from electrochemical brain processes, not from an immaterial soul that operates in ways no instrument can discover. . . . Views of human nature rooted in humanity's tribal past ought not to be our primary criterion for making moral decisions about cloning. . . . The potential benefits of cloning may be so immense that it would be a tragedy if ancient theological scruples should lead to a Luddite rejection of cloning.

In response, the University of Chicago ethicist Leon Kass (who served as chair of George W. Bush's President's Council on Bioethics) vigorously attacked this statement in terms that were reminiscent of both Ramsey and Jonas, two of Kass's intellectual mentors. In doing so, he evokes ideas of human uniqueness, moral responsibility, and the arrogance of modern scientists:

> To justify ongoing research, these intellectuals are willing to shed not only traditional religious views, but all views of human distinc-

> tiveness and special dignity, their own included. They are seemingly unaware that the scientific view of man they celebrate does more than insult our vanity. It undermines our self-conception as free, thoughtful, and responsible beings, worthy of respect because we alone among the animals have minds, hearts, and aspirations that aim far higher than mere life and the perpetuation of our genes. It undermines the beliefs that hold up our mores, practices, and institutions, not excluding science itself. . . . The problem may lie not so much with the scientific findings themselves but with the shallow philosophy that recognizes no other truths but these and with the arrogant pronouncements of the bioprophets.

Kass goes on to criticize the Harvard psychologist Steven Pinker, who had written in a letter to the *New York Times* that "brain science has shown that the mind is what the brain does," thus refuting the idea of a human soul. "One hardly knows," Kass remarked, "whether to be more impressed with the height of Pinker's arrogance or with the depth of his shallowness." In a reply to Kass, Pinker wrote that his "talent for moralistic invective is apparently not matched by a commitment to scholarly due diligence," denying that he would reduce the mind to mere matter but restating the view that the soul is at best an unhelpful notion for any scientific inquiry. Continuing the tiff, Pinker published an article in the *New Republic* called "The Stupidity of Dignity" in 2008, in which he expresses the suspicion that the term *dignity* is freighted with certain religious sentiments and that its use by cultural conservatives, including embryonic stem cell research opponents, is an illicit attempt to import those values into the ethical discussion. Pinker's concluding broadside referred to "The sickness in theocon bioethics," Pinker's term for a bioethics rooted in religious conservatism, which he says,

> . . . goes beyond imposing a Catholic agenda on a secular democracy and using "dignity" to condemn anything that gives someone the creeps. Ever since the cloning of Dolly the sheep a decade ago, the panic sown by conservative bioethicists, amplified by a sensationalist press, has turned the public discussion of bioethics into a miasma of scientific illiteracy. . . . Conservative bioethicists presume to soothsay the outcome of the quintessentially unpredictable endeavor called scientific research.

So much for the usual academic courtesies. The remarkable verbal brawl between two such distinguished scholars illustrates how much the perceived stakes have risen as the subject matter of bioethics has been transformed into that of the new biopolitics.

Who's in Charge?

Regardless of the accuracy of Hans Jonas's notions of the implications of genetic control, he was prescient in this more general sense: social institutions will have to prove that they are legitimate stewards of the power of the new biology. At this early stage, mistrust of governance institutions, from the state to industry to the science establishment, abounds. As a result, the biopolitics now unfolding is in some respects a version of what German philosopher Jürgen Habermas calls a "legitimation crisis." For Habermas, a legitimation crisis occurs with the impotence of government oversight and subsequent loss of public confidence in government's ability to manage public affairs. The legitimation crisis that is now energizing the new biopolitics is an accelerating crisis not only of public trust in government's ability to manage the science, but also of the public's persistent ambivalence about the goals, power, and values of science and scientists. This problem is aggravated by the widespread sense that science is sometimes oversold, that promises are made that can't be kept—with the promise of the genome or the payoffs from stem cell research common examples—or that contradictory assertions pop up, especially with regard to subjects of universal and immediate concern such as diet and nutrition.

Consider our understanding of nature in its most general sense. Besides the growth of knowledge itself, the momentum behind the new politics of biology is being fed by a variety of factors, including the burgeoning global nature of science, the rapid growth of widely distributed expertise that globalized scientific cooperation generates, competition among countries for control of the knowledge and its application, jockeying between governments and private companies for access to information that will generate new wealth and power, and, especially salient, the question of whether certain aspects of the science should be pursued at all because of the possibilities for fundamentally altering society, human beings, and the more or less established array of all living things.

Moderns tend to look to government first for control of new technologies and their effects. Competent oversight of potential threats to personal security and well-being is one measure of public confidence in government; yet if it succeeds too well, government may also be suspected of excessive control over, and appropriation of, science. Government agencies, like the Food and Drug Administration (FDA), that seek to impose their jurisdiction over emerging science for the public interest may be accused both of overregulating, and thus discouraging innovation, and also of "knowing too much" about proprietary matters. On the other hand, conspiracy theorists and some human rights advocates fear that these agencies' missions arrogate too much information into the hands of shadowy bureaucrats. Less noticed, there is also the competing prospect of private control over basic knowledge about life, as in the case of genetic-testing companies that market analysis of one's genome and in the process collect vast amounts of genetic information.

In 2010, the FDA considered a proposal for the first genetically modified fish that is directly consumed by people. This happened when a company called AquaBounty put a gene for growth hormone from a chinook salmon species into Atlantic salmon, as the chinooks grow much faster than the Atlantics. The company's genetic engineers also added a gene from a fish called an ocean pout to keep the growth hormone working all the time. Everyone agrees that there is a salmon shortage, one aggravated in recent years by the global popularity of sushi, so this seems a brilliant solution. Tuna may well be the next fishy target. Critics complained that the genetically modified fish should be labeled as such; others pointed out that agricultural products such as corn and tomatoes have been modified for a long time, so labeling the salmon would be more misleading than helpful. The core issue here is trust in science and scientists and those who would profit from and regulate the science. Who can have the authority to decide about a future modified-life world? In an act of acquacultural biopolitics, the salmon proposal was blocked by a conservative member of congress.

Neither government nor industry is immune from the life sciences' legitimation crisis. Neither are scientists themselves. The scientific community's role in the direction of science has become a joker in the deck. As early as 1624, Francis Bacon's *New Atlantis* described a utopia in which a clique of scientists ensures an enlightened society through the continuous study and application of new knowledge. In doing so, they keep their powerful discoveries close to their chests, having judged them too dangerous to share with those not in the fraternity. Secret societies are nothing new; recall for example the ancient medical cult named for Hippocrates, whose members swore an oath of loyalty to their colleagues and teachers saying that they would keep their knowledge within the fraternal circle. Although their intentions may be benign, the fact that these in-groups often identify with some special knowledge makes those not within the circle nervous—particularly if that knowledge is regarded as a key to great power that conflicts with the natural order.

At the very outset of the Enlightenment, Bacon's vision set a pattern of joyous and unlimited discovery for the sake of human flourishing along with a narrative of special insight and control by scientists. Evidently, Bacon's governing scientists did not provoke suspicion among the other residents of the New Atlantis; or if they did, we didn't hear about it. The same is not always true of the often ambivalent relationship between modern scientists and their lay countrymen. In our own time, scientists are admired for their acumen (Americans always rate scientists among the most admired professionals) and the improvements to which their work may lead are much desired, but their increasingly technical and highly specialized vocabulary makes their discourse inevitably somewhat exclusive. Scientists' very ways of thinking and perceiving as they dig into underlying and invisible processes seem different from the ways we approach the obvious world of everyday experience.

Sigmund Freud's idea of the uncanny, a situation that is both familiar and disconcertingly unfamiliar, applies well to the fruits of modern scientists' thinking. This form of thinking routinely transforms understanding of the seemingly familiar—"solid" objects and fixed species, for instance—into the uneasily unfamiliar, as though we've never really seen them at all. Of course, scientists are not the only sources of the uncanny. There are false conjurers, alchemists, fiction writers, poets, and painters who often brilliantly challenge our sense of the familiar. But those who subject their observations to the crucible of experiment and demonstration, whose revisions of the world lead to nuclear weapons, genetically modified fish and rodents, and even the (so far) beloved but ubiquitous World Wide Web, may understandably elicit special scrutiny. (Remember when we found "webs" creepy?) They present fertile ground for accusations of hubristic overreaching, aggravated by the fact that, like all experts, they seem to talk in code. All the transparency in the world still may not easily enable many of us to understand what a geneticist or microbiologist or stem cell biologist or computer scientist is talking about when they converse among themselves. Even if we are inclined to trust their good intentions (as many are not), we may wonder who is keeping track of what they're doing and where it's all heading.

The "Invisible College" Redux

At almost exactly the same time that Bacon's scientist-governed utopian tract was published, in the real world of the seventeenth century, an exciting and fertile process of intellectual exchange brought together a network of innovative scholars. They included such notable figures as Robert Boyle, known for his law of gas pressure and volume; John Wallis, who helped devise calculus; Robert Hooke, who first described the biological cell; and Christopher Wren, a polymath of architecture, astronomy, geometry, and mathematics. The postal system of the day enabled them to participate in a kind of virtual seminar. Reaching across Europe, they sent letters to one another describing experiments and exchanged books with scribbled marginalia. Independent of brick-and-mortar establishment institutions, they were a collaborative network of experimenters composing a virtual institution that Boyle described as an "invisible college": invisible, that is, to church and state authorities. In a sense, the Internet's traditional online science journals and the more novel "wiki" sites that allow an unlimited number of collaborators to edit Web pages are a vast elaboration of the original invisible college model.

As of yet, what Caroline Wagner has called the "new invisible college" has not garnered substantial public appreciation. While there is now little support for the growing dynamism and independence of science as practiced in the global college, no doubt the trends behind it will intensify, facilitated by instantaneous and efficient modern communications technologies. Science now proceeds as a global political force of its own through what has been called a "world polity," resistant to local cultural sensitivities. Yet the world polity of science is itself closely connected to the triumph over the last several hundred years of the nation-state. In the words of one group of observers, "the global spread of the standardized nation-state is clearly connected to the worldwide triumph of scientific authority and the higher education system within which scientific authority is prominently displayed."

World output of science papers increased from 450,000 in 1980 to 1.5 million in 2009, with Asia surpassing North America for the first time in 2008. And the most productive "faculty" of the new invisible college may show up in surprising, even disconcerting, places. According to the same report, since 1990 Iran has grown eleven times faster than the global average in science production as measured by publication rate, not only in fields related to the nuclear industry but in the life sciences as well. In Turkey the growth rate was 5.5 times greater than the international average. Of course, these numbers don't necessarily signify the quality of the work, and these countries are still far behind the established centers of science. Yet there is no denying that the growth rate of peer-reviewed publications from Asia is an indication of the international and collaborative nature of knowledge production in the twenty-first century, which wouldn't have been possible without the new communications technologies.

Science as an organized practice has undergone change in other ways than the speed of information exchange in the past couple of decades. Now biologists trade in data as much as they do organisms. Vast growth in gene databases, the complexity of the data, and far more sophisticated computing tools are key elements of the new life sciences. As one scientist wrote in 2006:

> The past decade has completely changed the face of biology. The image of a biologist passionately chasing butterflies in the wilderness of an Amazon rainforest or losing sight spending hours staring in a microscope has been substituted by pictures of factory-like sequencing facilities and high-throughput automated experimental complexes. The technology has changed the entire fabric of biology from a science of lonely enthusiasts to a data-intensive science of large projects involving teams of specialists in various branches of life sciences spread between multiple institutions.

Far more than in the original invisible college, the ability of the scientific community to direct the course of knowledge production and to influence the nature and extent of government or industry control are facilitated by how modern life sciences are practiced. Rather than large, nationalized and factorylike systems that characterized the development of big physics projects, with the atomic bomb effort the paradigm case, today the most sophisticated life scientists tend to work in small, agile teams that collaborate through both physical travel and virtual teamwork via the Web in a globalized network of shared expertise and experience. In the words of a university geneticist: "Essential for the success of large biological projects is further development of collaborative environments that will allow the scientists residing in different locations and sometimes even on different continents to analyze, discuss, annotate, and view the data."

In one sense this is a transparent system, but the transparency is only effective insofar as one is part of the culture of the system. The elite network of knowledge exchange and discovery could exacerbate the lingering alienation between scientists and society. Yet elite control of science will not go unchallenged, and some of the same information technologies that have made global scientific collaboration possible will be sources of these challenges. Dissident researchers and devoted amateurs create Web sites and blogs that develop contrarian points of view. The generalized access to science data may also be harnessed by scientific establishments, as in the case of "prediction markets" that may be used, for example, to enable massive speculation on the most promising scientific hypothesis. Off-laboratory betting could be offtrack betting for wonky risk-takers with disposable income.

Trust in Science

The legitimation crisis about control over the life sciences is a bit different from public skepticism about the scientific consensus on certain questions. American scientists and educators in particular have long bemoaned pockets of resistance to evolutionary theory. Meanwhile, recent surveys show that over the past few years, the more people have heard about climate change, the less they believe it is happening. A single ill-phrased e-mail message from one environmental scientist to another referring to statistical "tricks" in the confirmation of climate change caused a firestorm when revealed in the press, and perhaps did lasting damage to sensible climate policy. Climate change is not the only example of public doubt about scientific consensus. In 2010 a University of Maryland research team found that when people are told that scientists doubt that ESP is true, they were more likely to accept claims about paranormal phenomena.

The standard strategy to mitigate popular mistrust of science, often voiced from the scientific community, is that the public needs more science education. If people only understood science better, it is argued, they would not be so skeptical of the nature of scientific evidence and the often tentative conclusions that are reached. But evidence suggests that those who worry most about the power of science, especially biology, follow developments quite closely; more exposure might serve only to give them more reservations about the implications of scientific research. Even if the skepticism about particular scientific claims were to disappear tomorrow, the motive force behind the new politics of biology would remain. A more comprehensive and promising mitigation strategy includes not only educating the public about science, but also educating scientists about public concerns. Scientists, many observe, don't know how to talk about their work to the lay public, and some don't want to invest the effort. The twin goals of the mitigation strategy—public education combined with more sympathetic attitudes toward the public from the scientists' side—deserve support. But the best public communications strategy cannot change the underlying reality that greater biological knowledge and its applications will have profound implications for our society. This fact is a key driver of the new biopolitics.

The information revolution is key to the way science is done in another sense. When life scientists do communicate, they are increasingly talking about biological systems that are reducible to or described as quanta of information. The most familiar example of biology as an information system is the DNA chromosome. The methodology of reductionism is now being put to work in bioengineering fields like synthetic biology. Information-based science is both a product of and contributor to another important phenomenon: the convergence of biology and chemistry. A prominent example is the use of a type of enzyme called a kinase to transfer inorganic chemicals from one molecule to another in order to create industrial or agricultural products. This system is now commonly used in laboratories to target small molecules like those involved in cancer. The space between the life sciences and physical sciences is also narrowing, as nanoengineers and computer scientists are learning how to use physical and informational science tools to reprogram human cells. Information-based, convergent, and global science is moving ahead at a furious pace. It would be shocking if all that potent activity, conducted by a network of experts working in a highly specialized language, did not generate concern.

I have alluded to the most vivid recent instance of these tensions, the debate about cloning and human embryonic stem cell research. This signal incident carried echoes of a longstanding if vague worry that in spite of scientists' good intentions, someday powerful forces could use the knowledge being produced to control biological nature to enforce their will—or even that some are conspiring to do so now. For good historical reasons, the forces in question have mainly been attributed to the state. But when dealing with the new science, the state isn't what it used to be. Although the modern state is unlikely to wither away as Marx predicted, neither is it clear that it will be able to exercise control over the direction of science as it once did, much less the ways that the public perceives the culture of scientists. As I've noted, even if laboratory operations themselves cannot be moved from one jurisdiction to another, Web-facilitated collaborations can distribute work in geographically dispersed networks, and virtual laboratories or "collaboratories" that are wholly computer-based in decentralized networks will be able to aggregate vast amounts of data and conduct more and more research using sophisticated models of natural processes. The information and research-based documents will reside in "cloud" computing systems to which experts will have access but whose information technology infrastructure they will not control. Already, consumer-based services are developing the architecture for these systems.

To be sure, there are numerous competing forces at work. The state is more of a player in science than ever, as "basic" research with no obvious or immediate applications remains largely dependent on government largesse. However, even when resources are provided by the state through grants-in-aid or contracts, much of the work is not observable, nor are its results predictable. Successful laboratories funded for one purpose tend to be asking multiple related questions simultaneously. So they may in the course of their work address relevant but somewhat different questions from those central to the grant application or even ones incompatible with corporate goals. They may also choose to publish results that are disconcerting to government funders; in extreme cases, labs have stumbled into or deliberately chosen to investigate questions that are regarded as threats to public safety. Dangerous pathogens have been produced in the course of addressing valid and important problems. Despite their best efforts at audit and review, no funding source—government, industry, venture capital, or philanthropy—has complete control over what happens in the lab, a fact of which they are aware.

Similarly, government has funded a plethora of laboratories to do defensive research in response to worries about bioterrorism, but the materials and training provided by these labs can later spin off into unpredictable directions. The historical concurrence of the emergence of the modern territorial nation-state and the growth of science is not wholly accidental. A major part of this relationship consists of governments' recognition of certain military necessities, which have spurred strategic investment in critical technologies. In these cases, defense usually needs to play catch-up to offense, as in the case of "cyber-warfare" right now. The same is surely true in the life sciences; microbiology has become an engineering as well as an observational discipline, leading to more threatening bioweapons possibilities against which defenses, such as new and safer smallpox vaccines, may be prudent.

The academics in the new field of bioethics charted all of these themes in the 1970s. In spite of deep disagreements about the proper moral course of the new biology, they agreed on its implications. What they did not foresee was the extent to which these esoteric philosophical issues would shape the wider political debate only several decades later. As if any more evidence for the full emergence of biopolitics was needed, when former president George W. Bush published his memoir *Decision Points* in 2010, he devoted an entire chapter to his position on embryonic stem cell research.

CHAPTER FOUR

THE STEM CELL DEBATE

SHORTLY AFTER PRESIDENT GEORGE W. BUSH'S REELECTION IN 2004, the then-chair of the President's Council on Bioethics led a group that circulated a "conservative bioethics agenda" on Capitol Hill. Their goal was to influence congress to take action on some of the new biotechnologies through what they called "a bold and plausible 'offensive' bioethics agenda," instead of a congressional approach that was "too narrowly focused and insufficiently ambitious. . . . We have today an administration and a congress as friendly to human life and human dignity as we are likely to have for many years to come. It would be tragic if we failed to take advantage of this rare opportunity to enact significant bans on some of the most egregious biotechnical practices." The group's work was widely noted, including in a report by the *Washington Post*.

Leon Kass was not acting in his official capacity as the chair of a presidential advisory board, but as an engaged citizen. Nonetheless, his role as the promoter of a legislative policy agenda was, at the very least, a highly unusual one. It was the first of its kind by the leader of a sitting bioethics commission. While the effort undoubtedly rallied conservative elites, it also served to confirm the worst fears of many scientists that they faced an administration unfriendly to cutting-edge biology. Whether or not they were right about the administration's intentions, the conservative bioethics document was a sentinel event. In the context of years of national controversy about research on embryonic stem cells, it was another signal of the full flowering of the new biopolitics. Even in a new field like embryonic stem cell research, where the science is exciting but the medical benefits mostly promissory, the combination of human embryo use with increasingly powerful biology produced a symbolism that galvanized the political system. It was a system that had also entered an era of ideological extremes fueled by a twenty-four-hour news cycle.

Let the Debate Begin

The stem cell debate began in 1998, when researchers from the University of Wisconsin and Johns Hopkins University isolated the first stem cells from a human embryo. The issue surfaced in three presidential campaigns, including 2008, when both major presidential candidates declared their support for the research. The unfolding controversy started with an arcane lab technique and an esoteric bioethical debate that occupied a small circle of ethicists, expanding into an international cultural and biopolitical rumble that attracted polemicists, advocates, and finally politicians. Although the process of "deriving" stem cells from embryos left over in fertility clinics was highly technical, it called attention to the question of the embryo's moral status in relation to progress in medical science, creating what many saw as a conflict between two deeply held values. For scientists and other advocates, the potential benefits of understanding how these "pluripotent" cells become all the hundreds of cell types in the human body would not only provide stunning insights into early life, but would also ultimately help lead to an entirely new kind of "regenerative" medicine in which we use our own cells to heal ourselves. But others saw the use of the cells as an offense against the source of all human persons and another step down a perilous slippery slope. Who could, and who would, control this power over our early cellular development?

The embryonic stem cell debate was and is unique in an important respect: it is a scientific enterprise involving the direct destruction of human embryos. Previous controversies involving human reproduction also raised the "life" issue, if not quite as directly or emphatically as stem cells, and they had their own biopolitical undertones. One notable incident was the birth of Louise Brown in 1978 by means of in vitro fertilization (IVF), after which some heralded an era of lab-created monsters or the arrival of a brave new world. Because the results of IVF have come to seem benign, with millions of IVF births and Ms. Brown herself conceiving and having a baby the old-fashioned way, the anxiety that accompanied the event is all but forgotten. Still, that brief but vigorous discussion prefigured much of the later debate about other artificial reproductive technologies, such as preimplantation genetic diagnosis or PGD (screening embryos in a lab dish for various genes that would subject any resulting baby to a high risk of disease) and reproductive cloning (the process that produced Dolly the sheep in 1996 and many other large mammals since then).

In the run up to the full flowering of the new American biopolitics there were other events that involved newborns rather than embryos and fetuses. In 1984, the Reagan administration weighed in on a series of highly publicized "Baby Doe" cases, in which parents decided to withhold treatment from gravely ill newborns. The Surgeon General and other administration officials worried that a new code word for eugenics was being framed in terms of "quality of life" judgments. The U.S. Department of Health and Human Services tried to apply federal law prohibiting discrimination against the handicapped to decisions about life-sustaining treatment for severely ill newborns. The constraints imposed on neonatal intensive care nurseries included a notorious "hot line," a 1-800 number advertised on large posters in hospital nurseries for anyone to voice complaints about discrimination against sick babies. This heavy-handed approach didn't survive public or legal scrutiny, but it did signify a growing awareness that the care of seriously ill newborns had dramatically advanced in the two decades since 1962, when President and Mrs. Kennedy's son died after a five-and-a-half week premature birth: such an outcome would be unheard of in virtually any advanced country today, even for the least advantaged mothers and babies. But the problem that became clear in the 1980s persists. Those lifesaving advances could not always prevent medical problems that would then require lifelong dependence on complicated technologies and devoted caregivers.

In still another incident, an attempt to create a bioethics commission composed of members of Congress along partisan lines ran aground in the late 1980s. Unable to balance representation on the abortion issue, Congress could never agree on the precise membership and the body never met. In 1989, the legislation that created it ran out its clock. Perhaps the penultimate step in the bio-politicization of ethics took place in 1994, when a federal panel on human embryo research protections recommended government funding of research on human embryos because it could provide important benefits for infertility treatments, genetic disorders, and many human diseases. The panel also concluded that in some cases, embryos could be created for the sole purpose of research. In a response that echoed through the later stem cell debate, President Clinton rejected the recommendations of his own advisory panel, determining that no federal funds could be used to create human embryos that would then be destroyed in research.

In 1995, Congress attached language to the National Institutes of Health (NIH) budget that turned into law the prohibition against making embryos with federal funding. The Dickey-Wicker amendment was named for the two members of Congress who introduced it. When it was passed, no one anticipated the question that would be raised by the creation of human embryonic stem cells only three years later. Namely, could the NIH fund studies of cells that came from those leftover embryos? In deciding in August 2001 that the NIH could give grants to scientists who did research on cell lines derived from human embryos, President Bush implicitly accepted a 1999 determination by the chief NIH lawyer under President Clinton that stated research on cells was allowable under the law, though the cellular derivation process that results in an embryo's destruction was not. In effect, the edifice of federal funding rested on a single memo by a government lawyer. Was that enough?

Fifteen years later, that question would precipitate a legal crisis for the future of embryonic stem cell research in America.

Enter the Clones

First, though, came Dolly. The scientists at the University of Edinburgh's Roslin Institute were not seeking controversy when they developed a technique for removing the DNA from an adult ewe, inserting it into a hollowed-out egg, and stimulating the product to divide and grow much like a normal embryo. Rather, they were trying to find a way more efficient than traditional breeding for the production of the most useful farm animals. But, as it so often does in the new biology, controversy found them. After hundreds of failures due to anomalous fetuses and danger to the animals into which the activated eggs were implanted, Dolly was born in 1996. In a bit of wry Scottish wit (Edinburgh did after all produce stellar intellectuals like Adam Smith, Sir Walter Scott, and David Hume), she was named after Dolly Parton, owing to the source of her DNA, a mammary gland.

As the first cloned mammal, Dolly made real one of the worries of many philosophers and theologians in the 1970s. This was the fear that cloning would be among an array of technologies to "manufacture" babies with certain characteristics. Not surprisingly, Joseph Fletcher thought that cloning would be an expansion of human freedom and far superior to the usual "genetic roulette," going so far as to call the procedure "radically human" because it would represent vastly more deliberate reproduction. Paul Ramsey, consistent with his other views about the new biology, called cloning a "borderline" technology that pushed up against moral limits, leading to managed breeding and experiments on the unborn while also threatening the meaning of parenting as mere reproducing and disconnecting sex from parenthood. To Ramsey, cloning also represented another instance of scientists' hubris in which humans strive to become gods.

As always, movies and novels like *The Boys from Brazil*, in which dozens of little Hitlers are cloned to take over the world, primed public perceptions. Some even seemed to think that a clone would have the same mind or soul as the original. What many failed to realize is that there have always been "clones" of other individual, identical twins. In fact, twins that result from the spontaneous division of a single embryo share more of their DNA than clones do, because identical twins also have the same DNA in the egg's mitochondria. But the symbolism of the cloning process raised alarms. President Clinton immediately placed a ban on human cloning and ordered his just-appointed National Bioethics Advisory Commission to study the ethics of cloning. The commission concluded that cloning for reproduction presents unacceptable medical risks to both mother and fetus and might create psychological problems for the child who survives the procedure. Clinton agreed.

Over the next ten years, all kinds of political twists and turns accompanied the human cloning issue. Yet the legality of cloning for reproduction remains largely unresolved, except in a few states. When congressional Democrats introduced a bill in 2007 that would have made reproductive cloning a crime, conservatives opposed it because it did not prohibit cloning for research. Although the Bush administration was, by all accounts, no fan of the United Nations, it chose that body as the forum for an effort to achieve an international convention against human cloning for any purpose. The effort fell short. Instead, in 2005 the United States settled for a less binding "declaration" on human cloning, which was passed by a majority of nations. The member countries that opposed the General Assembly resolution, including the United Kingdom, noted that it could be read as opposing both cloning for research as well as cloning for reproduction.

Achieving international agreement on cloning has been difficult. Moreover, when it has been achieved, there are technical problems. Legislative language, for instance, whether in a national parliament or in transnational agreements, can inadvertently compromise science. In 1998 the Council of Europe adopted a protocol, "Human Rights and Dignity of the Human Being with Regard to the Application of Biology and Medicine, on the Prohibition of Cloning Human Beings." The Council protocol proscribes "any intervention that seeks to create a human being genetically identical to another human being, whether living or dead." But this rule actually bans the creation of identical twins through IVF (since twins are "genetically identical"). Furthermore, it doesn't rule out cloning because, unlike a true identical twin, a clone lacks the mitochondrial DNA of its source individual. That genetic material, often called the cell's power plant, is located outside the nucleus.

Stemming Progress?

Dolly's significance had barely been digested when the news broke in 1998 that a University of Wisconsin lab had isolated human embryonic stem cells. For decades, people had theorized that there must be some "master cell" in the early embryo that gives rise to the two hundred-plus cell types in the human body. Then, cells that were made from mouse embryos suggested that there must be such cells in humans. Now we had that master cell. If the cell turned into all those tissues as the embryo develops into a fetus, we potentially had at hand the key to using our own biological materials to heal ourselves by replacing defective tissues in hearts, livers, pancreases, brains, or in fact any other organ or tissue system in the body. In a way, science would be recapturing the ability to regrow damaged organs have largely lost in evolution, but which many reptiles have retained.

Since cloning is one way to obtain embryonic stem cells (another way being from an embryo as nature understands it, through the fertilization of an egg by a sperm), the first tidal wave of publicity about cloning and Dolly merged into the second about embryonic stem cells. The paper from the Wisconsin lab finally made it impossible for many formerly academic bioethical questions to be isolated from embryo politics, for science now seemed to require the destruction of embryos in order to pursue a highly promising research field. The scientific promise of human embryonic stem cell research, when set against the question of the embryo's moral status, acted as a wedge in fissures that had scarred over. It also contributed to the decisive and perhaps irreversible transformation of bioethics into biopolitics, the politicization of the ethics of biology.

From its inception, the new biopolitics threw a wrench into the predictable left-right political framework. Consistent with his position on human embryo creation, President Clinton opposed using federal funds for reproductive cloning but agreed with his bioethics commission that stem cells could be obtained from embryos left over in fertility clinics. Thus, President Clinton had many of the same reservations about cloning and stem cell research as President Bush. However, unlike Bush, Clinton showed no inclination to ban federal funding of research on embryos that were donated with informed consent and left over in fertility clinics. Later, conservative senators, John McCain and Orrin Hatch among them, supported using cells from human embryos from fertility clinics in research, at considerable cost to their support from social conservatives.

The coinciding developments of cloning and stem cell research thrust the embryo's place in science into the public arena as never before. Pro-life organizations sought the attention of the 2000 Republican presidential candidates, including then-Governor George W. Bush. Though the issue barely broke the surface during the 2000 campaign, soon after taking office President Bush announced that he was reviewing the Clinton policy. Then, in the speech from his Texas ranch, President Bush explained what he was allowing: federal funding only for research on embryonic stem cell lines derived from human embryos before that night. In this way, the president was able to split the difference, enabling promising research to go forward while taking a stand against research that encouraged more embryo destruction. But, as Supreme Court justice Sandra Day O'Connor said of abortion, this was a policy at war with itself, because any success with embryonic stem cells would put pressure on the federal government to change its policy. In the event, the Bush administration could leave that decision to its successors.

Behind the Stem Cell Lines

After the Bush policy announcement, scientists who believed that work on cells from embryos donated with consent was justified were left with unsettling ethical and regulatory worries—even if that research was limited to leftover embryos. At first, there was as much relief among biologists as there was consternation among pro-life advocates. Many scientists had expected the president, whose personal opposition to abortion was well known, to ban the funding altogether. Their relief was bolstered the next day, when a press conference held by NIH leaders announced that there seemed to be dozens of human embryonic stem cell lines around the world that could be used in research under the policy. However, over the next few days and weeks, it became clear that this was a wild overestimate. Finally, there were fewer than two dozen such collections of cells. Slowly the scientists' relief turned to anxiety, as it was uncertain what the quality of lines was, how long they would be truly viable, what the ownership and patent restrictions were, and if they were genetically different enough to allow for a broad range of new knowledge.

The regulatory worries were in some ways even more vexing. The Bush policy only applied to embryonic stem cell research supported by the federal government. So what were the rules governing embryonic stem cell research done with other funding sources, like private industry and the states? What was the Food and Drug Administration's role? By law, it is ultimately the FDA that has to approve biological materials for medical use if they are more than "minimally manipulated." And what about other science codes of ethics? How did they apply? Most scientific research on human beings or human materials requires transparency and various levels of review. Since the Bush policy did not directly restrict research that used nonfederal funding, the scientists wanted guidance on stem cell work using state or private resources.

Chartered by congress as the nation's advisers on science, engineering, and medicine, the National Academies are loath to step into charged political territory. Much of their prestige and credibility, in fact, is based on theirr reserve about such matters. In 2004, the National Academies agreed to appoint a committee to develop research guidelines for embryonic stem cells. The Academies had already published a report concluding that at least some kinds of research involving human embryonic stem cells were of scientific importance and should be permitted, but these did not include reproductive cloning. The new committee was charged with elucidating the conditions under which research might acceptably take place. After more than a year of meetings, in March 2005 the Academies released the voluntary guidelines. The main points were that any reproductive materials for embryonic stem cell research, including sperm, eggs, and embryos themselves, must be given with consent and without payment. Also, research institutions should create systems that will enforce ethical conduct of the research.

The stem cell report made national headlines, including an above-the-fold front page story in the *New York Times*. Perhaps the principal immediate result was reassuring congress that the scientific community was prepared to establish its own voluntary standards in the absence of detailed guidance from government. In this respect, the report could be compared to the voluntary moratorium on DNA research that came out of a meeting of biologists at Asilomar, California, in 1975. Virtually every major research institution and science organization immediately announced that they would adopt the Academies' recommendations. A number of other countries, including China and India, used the guidelines to help develop their own policies. California, where voters had approved three billion dollars of state funding for stem cell research, integrated a version of the guidelines into state law. The new International Society for Stem Cell Research developed its own guidelines, which were largely consistent with those of the Academies.

Over the next few months a bill to open up more stem cell lines for research gained supporters. Representative Diana DeGette, a Democrat from Colorado, and Representative Mike Castle, a Republican from Delaware, were the co-sponsors in the House of Representatives. Senators also crossed the aisle behind the measure, led by Utah's pro-life Republican Orrin Hatch and the Democratic Senator Tom Harkin of

Iowa. In hearings, the divide between the administration and its own NIH director was evident when Dr. Elias Zerhouni said the Bush policy required scientists to work with one hand tied behind their backs. The bill was passed by both houses of Congress, only to be vetoed by President Bush—his first veto since taking office—in a duet that was repeated a year later.

Just as the Obama administration was rolling out new embryonic stem cell lines for approved NIH-funded research in 2010, a court case revived the cultural conflicts. Two scientists who work on adult stem cells joined with pro-life organizations in a court challenge of the interpretation of the Dickey-Wicker Amendment that had been in effect in the Clinton, Bush, and Obama administrations. That interpretation, based on a 1999 interpretation by a government lawyer, allowed the NIH to pay for research on stem cells from embryos. The plaintiffs argued that the amendment banned funding even of the cells obtained from embryos, not only their direct destruction. The immediate issue was whether the standing interpretation of the federal law was a reasonable one or simply an instance of clever lawyerly hair-splitting to reach a desired goal. It was the first time that the legal system had been the venue for a challenge by opponents of embryonic stem cell research.

The pending legal case flew under the radar of the life sciences establishment, which had thought that, as far as the stem cell issue was concerned, the worst was behind it. When the case broke, many stem cell research advocates came to feel that in retrospect, it was a mistake to rely on a single legal memorandum for such a controversial public issue. Biopolitics has a life of its own that is not always amenable to bureaucratic solutions.

Bioethics in the New Biopolitics

Along with his stem cell policy, President Bush also created a bioethics panel, the President's Council on Bioethics, that was charged with reflecting on "ethical issues connected with specific technological activities, such as embryo and stem cell research, assisted reproduction, cloning, uses of knowledge and techniques derived from human genetics or the neurosciences, and end of life issues." Unlike most presidential commissions, including those on bioethics, it was not required to reach consensus. Instead, its charter focused on attaining a "a deep and comprehensive understanding of the issues that it considers." The council's composition immediately excited concern from biologists who worried about the conservative intellectual orientation of many members. It was chaired by Kass and also included the bright and acerbic *Washington Post* columnist Charles Krauthammer. A philosophical viewpoint that emphasized the need to not upset long-standing cultural and moral traditions was dominant on the council. But there were also several prominent scientists who did not have a particular record of political engagement, thus diluting the criticism that the President's Council was a stacked deck. These scientist members, however, made no secret of their concern about the way some of their fellow council members viewed the new biology.

Biologists' suspicions about the council's attitude toward science were further aroused a couple of years later when two members were not renewed after their terms expired. One member was the distinguished, politically moderate theologian William May, the other was prominent stem cell biologist Elizabeth Blackburn. In 2009, Blackburn won the Nobel Prize for her contributions to stem cell research, providing sweet vindication for the Kass panel's critics. Although no reason for the non-renewals was ever publicly stated, from a public relations standpoint the episode was disastrous for the council, giving many in the scientific community yet another reason to hold its motives in contempt and associate it with what science writer Chris Mooney called, in his 2005 book of the same name, "the Republican war on science."

The council membership issue aroused the worst fears of science advocacy groups, especially with regard to the role of Blackburn. Ultimately, the episode brought many life scientists into vigorous political activity for the first time during the 2008 presidential campaign, as they rallied around candidates they saw as friendlier to science. By the time Kass left the Bush bioethics council chair, a great many life scientists had been deeply alienated by what they perceived as a moralistic and judgmental attitude from the modern conservative movement. Important biologists were nearly apoplectic about what they saw as a threat to scientific freedom presented by the Bush administration. Especially at a time when the conservative Republican tides seemed on the brink of accomplishing permanent majority status, many wondered if a deeply moralistic attitude would come to characterize future government science funding. The several government scientists and physicians who reported that their work had been compromised by political commissars added to the disquiet. So did controversies that were, fairly or not, identified with the Bush administration's politics, such as that about "intelligent design." By 2008, many life scientists began to feel that they were witnessing the birth of a new culture war about science, the stem cell issue being the canary in the coal mine.

Embryo Ethics

In addition to the membership controversy, just as revealing about the philosophical orientation of many on the Bush bioethics council was its chosen agenda. Much of its discussion and writings focused on the question of human dignity in the context of biotechnology. Considering that a presidential commission saw fit to give the topic special attention, an observer might have reasonably inferred that respect for humanity itself is at grave risk from modern applied biology. Imagine instead that the council had chosen justice in health care as a theme. That choice would have lent some credibility to the hypothesis that the American health care system might well be unjust. By focusing on the way that the life sciences may undermine human dignity, the council not only articulated long-standing discomfort about science, but it also sent an implicit message that the threat is great enough to be considered at the highest level of government.

Claiming that critics were politically motivated, the Bush bioethics council's defenders often noted its pluralistic composition. They noted that President Clinton's bioethics commission didn't seem to include any "pro-life" advocates—or at least none who were obvious. The best-known progressive thinker on President Bush's bioethics council was Michael Sandel, a distinguished Harvard professor who was the best argument against the view that the council's membership was stacked. Importantly, though, Sandel has long emphasized the importance of context, community, and tradition in ethics, distinguishing himself from social contract theorists like John Rawls, a philosophical hero of many mainstream American liberals. In that sense, although a progressive, Sandel fit in well with the council's general orientation.

In 2004, while he was a member of President Bush's bioethics council, Sandel published a widely read piece in the *New England Journal of Medicine* on "Embryo Ethics." The immediate question was whether respect for the human embryo ruled out using it in research that might provide important medical advances. Sandel argued, in part, that those who ground respect for the human embryo in the fact that all persons came from embryos have in effect committed a logical error. "Consider an analogy," he wrote,

> [A]lthough every oak tree was once an acorn, it does not follow that acorns are oak trees, or that I should treat the loss of an acorn eaten by a squirrel in my front yard as the same kind of loss as the death of an oak tree felled by a storm. Despite their developmental continuity, acorns and oak trees are different kinds of things. So are human embryos and human beings. Sentient creatures make claims on us that nonsentient ones do not; beings capable of experience and consciousness make higher claims still. Human life develops by degrees.

In this passage, Sandel conflated at least two different arguments against the notion that the origins of human persons have the same moral status as persons. First, that this notion is an example of the "genetic fallacy," defined as an argument that judges the worth of something based solely on its origins rather than its current circumstances. Second, it commits what philosophers call a category mistake, ascribing qualities that the object in question could not have. Both are arguments based on logic and language.

Partly because it appeared in an important medical journal, Sandel's paper attracted the scientific community's attention far more than the typical philosophy article. On the other hand, the very fact that such a prestigious journal would publish such a piece confirmed the suspicions of cultural conservatives that the life science establishment just didn't "get" the moral concerns of many Americans. In addition, by addressing the moral status of the embryo so explicitly, the paper seemed to acknowledge a link between the stem cell issue and the abortion debate, even though embryonic stem cells are obtained from embryos in fertility labs, not in a uterus.

In response, Sandel's colleague on the bioethics council, Princeton professor Robert George, coauthored a paper in the conservative journal *The New Atlantis* in which he deprecated Sandel's analogy. George wrote that "Sandel's defense of embryo-killing on the basis of an analogy between embryos and acorns collapses the moment one brings into focus the profound difference between the basis on which we value oak trees and the basis on which we ascribe intrinsic value and dignity to human beings." We value mature oak trees for their aesthetic qualities and mourn them when they fall because of their "magnificence," not because of the kind of entities they are, George claimed. By contrast, he argued, we mourn the loss of an adult human being precisely because of the essential qualities of human beings, not because of incidental aesthetic trait like magnificence. Because embryos are human beings, we should value them on the same grounds as all other human beings, owing to their intrinsic character.

When intellectuals participate in biopolitical debates, language is sometimes just as important as evidence. George and his coauthor Patrick Lee adopt the apparently question-begging formulation that equivocates between "human beings" (which the human embryo is) and "persons" (which is exactly what is at issue). More to the point, however, the Sandel-George exchange epitomized the rhetorical disadvantage of the embryonic stem cell research advocates. In comparison to the language of their opponents, theirs often seemed cold and calculating. The opposition wanted to widen the scope of moral concern about human beings to include the human embryo. "The dignity of human beings," George and Lee wrote, "is *intrinsic* to the kind of entity we are; it does not depend on accidental attributes like size, skin color, age, or IQ." And who would want to reduce human dignity to such trivial traits? But there was a still more significant fact about this exchange that was unremarked upon at the time: it was the first instance in which a discussion between members of a presidential bioethics commission was brought squarely into a public debate. This provided still more evidence of the transformation of bioethics into biopolitics.

Not all the objections to human embryonic stem cell research were specifically ethical or tied to the alleged hubris of scientists. The philosophical reservations were complemented, and sometimes conflated, with conflicting claims about the science. The Bush administration and some members of his bioethics council were staunch advocates of less ethically charged sources of cells that could in theory do the work of embryonic stem cells. Sometimes it seemed that these goals were reducible to a theory in search of the facts. At other times, exaggerated and misleading claims were made about the prospects that so-called "adult stem cells," in use in medicine for decades, could be modified to become other cell types. These claims were combined with the oddly self-fulfilling proposition that the research should be stopped because "embryonic stem cells haven't cured anyone yet." But other sources of potent cells, such as those in amniotic fluid, continue to have promise. What was needed was a pluripotent cell, one that resembled the ability of cells from the human embryo to turn into hundreds of cell types. And until late 2007, embryos were the only source of these potent cells.

Beyond the Embryo

For years, stem cell biologists had theorized that there must be a way to "de-differentiate" cells that nature had programmed to become certain types. The aim was to make other cells as potent as those derived from human embryos. Top scientists in meetings I attended in the early 2000s estimated that it could take twenty years to accomplish this feat. Thus, when Japan's Shinya Yamanaka and then Wisconsin's James Thomson reported in separate 2007 papers that only a few genetic factors were required to de-differentiate an adult skin cell, there was universal delight and amazement. There was also an undeniable air of triumphalism among those who had opposed embryonic stem cell research. The White House had for months been eagerly anticipating the announcement about "induced pluripotent stem cells." When the moment came, both presidential spokesmen and allied political commentators declared that the use of human embryos for research had finally been superseded. At last, it seemed, they had science on their side.

In fact, some scientists did believe that human embryonic stem cell research was now an anachronism. With a bit more work, they thought, the pluripotent stem cell process would produce cells that were equivalent in every way to those derived from embryos. And the clunky, mechanical process of nuclear transfer or cloning as well as the controversies about a source of eggs would be left behind. But while a few scientists were indeed reorienting their labs to the new technique for producing pluripotent cells, most were not willing to abandon embryonic cells. They noted that any manipulation of skin cells with genes or chemicals, as well as the unmanageable conditions of a lab environment, would produce modifications in the cells that might never be wholly predictable. This could result both in cell lines that, though pluripotent, have properties unique to each line, and these manufactured cells might never be safe enough to put into human patients. It was reported that the new lab-created pluripotent cells had higher rates of cell death than embryonic stem cells but did not proliferate nearly as quickly. And the cells taken from adults seem to retain the "genetic memory" of the cell systems they came from even after they are returned to a pluripotent state. If these results are confirmed, then pluripotent stem cells from adult cells won't be as useful in therapies as those from embryos could be unless other "fixes" can repair the damage created by inducing pluripotency in adult cells and unless their genetic memory can be fully erased.

No one can be sure of the limits on laboratory manipulations imposed by biological laws that science has not run up against yet. So it is dangerous to predict the outcomes of such basic research or how techniques in one field might turn out to be useful in another. In a classic example of the "spin-off" phenomenon in science, the transfer of a nucleus from a cell to an egg has turned out to be a promising technique for helping parents have genetically related children without the risk of passing along diseases carried in certain genes. Some women are at high risk of having an infant with devastating neurological disorders transmitted by DNA in the egg's mitochondria, outside the nucleus. "Egg cell nuclear transfer" consists of inserting the intended mother's nucleus into a healthy donor egg. The egg with its new nucleic DNA can then be fertilized in the lab with her male partner's sperm, thus avoiding mitochondrial disease. Opponents of human cloning want to ban "reproductive cloning," but that language could inadvertently also prohibit this procedure, which many would find ethically acceptable. This is the problem with banning certain scientific technologies with the blunt instrument that is the law: the wrong language can have unintended consequences that are not easy to repair.

Whatever the ultimate usefulness or limitations of induced pluripotent cells, the insights that led to their production were built on a decade of research on the human embryonic stem cells, which remains the gold standard of pluripotency. Similarly, one of the measures of pluripotency for these artificially created cells is a comparison of their properties with those of human embryonic stem cells. The first American to produce pluripotent cells, James Thomson, himself endorsed continued embryonic stem cell research and attended President Obama's ceremonial signing of an executive order overturning the Bush policy, despite having acknowledged the ethical issues surrounding embryo destruction. Decrying the policy change as a betrayal of human embryonic life, opponents also noted Yamanaka's remark, at the time of his publication on induced pluripotent stem cells, that he was bothered by the use of human embryos and hoped he had provided an alternative.

Thus, the stem cell controversy took place at various levels and among various parties contending for control over stem cell science. While the public controversy about the moral status of embryos and the importance of medical science unfolded, a legislative debate continued on somewhat different premises. Advocates and opponents eagerly awaited scientific breakthroughs and hoped for encouraging words from scientists. There was an academic self-governance response and, waiting in the wings, government regulators in the Food and Drug Administration were quietly prepared to assert their authority over human and animal stem cell experiments. Still further in the background were those who were prepared to lay claim to the property rights over any valuable products of embryonic stem cell research. A notable example was the University of Wisconsin spin-off that claimed to own the patent on deriving stem cells from embryos.

Divisions between and within the political left and right showed that the matter was as much cultural as it was political. Traditionalists on the right fretted about threats to human dignity, while their more libertarian colleagues held reservations about undermining scientific innovation. Although liberals and progressives largely supported the biologists, many did so with lingering concern that the fruits of the research would once again serve to benefit the affluent and that stem cell research would turn out to be another chapter in a long march of compromised human dignity. When legislation passed in California that made that state a leader in funding embryonic stem cell research, Latina organizations were alarmed that state funds could pay poor women for risky egg-retrieval procedures. It was another case of biopolitics defying standard political alliances in the struggle for control over the new biology.

Stem Cells Abroad

The new biology is global, but biopolitics remains irreducibly local. Certain localities are especially poised to take advantage of the uncertainties about stem cell research in the United States. The stem cell issue provoked varied responses among the many countries in the hunt for the prestige and potential market value of embryonic stem cell breakthroughs, with the new communications and information-sharing technologies doing much to level the scientists' playing field. Those with strong computer and agricultural science systems had a strong start. Another advantage went to countries with a centralized industrial policy that sought to use biotechnology as a twenty-first century industrial platform, among them China, Singapore, and South Korea. An analysis by a University of Toronto group indicated that China published twenty times as many stem cell scientific papers in 2008 as it did in 2000. Chinese labs have produced at least twenty-five human embryonic stem cell lines and perhaps as many as seventy. The stem cell example is a window into Chinese advances in biology. China is now second in published papers on the biomedical sciences, according to the National Science Foundation's 2010 science indicators report. However, China also has problems with unregulated "stem cell tourism" due to clinics advertizing various unproven treatments to patients, especially Westerners desperate for cures.

The Korean experience developed into a special warning for those too eager to grasp the prestige and value of leadership in the new biology when a top Seoul University scientist claimed to have cloned embryonic stem cells. (He also was revealed to have paid a number of women fourteen hundred dollars each for their eggs.) Before his data were discovered to have been fabricated, he was declared a national hero, given a prestigious professorship, awarded free lifetime first-class seating on the national airline, and had a stamp cast in his honor. When his fraud was discovered, a period of national shame followed, as well as recriminations from some stem cell research opponents about the inevitability of overreaching by ambitious scientists. The scandal in South Korea fed the long-standing cultural conservative narrative about hubris.

On the whole, the countries with strong biology and non-Christian religious traditions have been aggressively pursuing all sorts of stem cell research, with little internal debate. Israel and, to a far lesser degree, Iran have been working on embryonic stem cells. It turns out that both Jewish and Islamic law are fairly permissive about research involving embryos if human lives can be saved. In Europe the picture is more mixed, reflecting the varying influence of the Vatican. With its Anglican tradition and centralized system for research on human fertility and embryology, the United Kingdom has awarded several licenses for human research cloning, establishing the nation as a leader. A number of countries prohibit creating embryos for experiments but allow them to be imported. One example is Germany, which does not permit the destruction of embryos in research or even for preimplantation genetic diagnosis, practices that are considered to be excluded by its antieugenic, postwar Basic Law. However, in a remarkable irony, German scientists do conduct research on human embryonic stem cell lines imported from Israel. These same lines were among those approved for use in the United States during the Bush administration.

Owning the Idea

Finally, another party in the competition for control over the new biology has been waiting in the wings, finding itself in an awkward position as the public controversy unfolded: the corporate sector. In the United, States, the main players are the University of Wisconsin Alumni Research Foundation (WARF), which claims patent rights over the process of deriving stem cells from human embryos, and two corporations, Geron and Advanced Cell Technology, that are trying to develop medically useful products from embryonic stem cells. WARF has spent quite a bit of time in court defending its right to license the lab process for stem cell derivation based on the 1998 work of UW professor James Thomson.

In 2010, Geron obtained approval from the Food and Drug Administration for a clinical safety trial of an embryonic stem cell treatment for patients with recent spinal cord injuries, a condition that received a lot of attention due to the injury that took the life of actor Christopher Reeve. Reeve and his wife, Dana, who succumbed to lung cancer less than two years later, were vigorous stem cell research advocates. Together with Michael J. Fox and Nancy Reagan, they brought celebrity power and helped put a human face to the pro–stem cell side. In the Geron experiment, patients will receive infusions of embryonic stem cells that have been differentiated into cells capable of producing myelin, the coating that conducts electrical impulses in the spine. There is no expectation that this cell treatment will magically regenerate spinal cords, though much could be learned from a greater understanding of how new cells integrate into damaged tissue. Rather, the goal is to facilitate, along with strenuous physical rehabilitation, some improved potential for movement. Advanced Cell Technology, hopes that its embryonic stem cell treatment will succeed in improving the outlook for young people with Stargardt's disease, a gradual loss of eyesight that starts in childhood and ends in blindness.

Intellectual property disputes can be tedious and enormously expensive. If they involve international claimants, they can take a decade or more to resolve, sometimes with the help of high-level government negotiations. Although few of us pay attention to these legal battles, they are the critical fronts in efforts to wrest wealth from innovation and shape the ways future industries are governed. The 1980 U.S. Supreme Court decision that made it possible to patent genes seemed pretty esoteric at the time but, along with the federal law for technology transfer from federal agencies to the private sector that same year, it changed the course of history in capturing the value of biotechnology.

The case, *Diamond v. Chakrabarty*, involved a patent for a genetically modified bacterium that could break down crude oil in the event of a spill. The court found that gene patents satisfy the requirements of the federal patent statute, which, with one exception, is the same as Jefferson's draft: "Whoever invents or discovers any new and useful process, machine, manufacture, or composition of matter, or any new and useful improvement thereof, may obtain a patent therefor, subject to the conditions and requirements of this title." All that has changed since Jefferson is the word "process," which was originally "art." The effect of this favorable judgment was that, in the United States at least, life forms could be patented and therefore in effect owned as property, a philosophical position that much of the rest of the world has resisted. Thus, the aggressive pro-technology orientation of the founders has come to shape ownership of twenty-first century biotechnology. Jefferson's apparent wish for the new nation to recognize and reward invention, perhaps beyond all other countries, continues to be fulfilled, but according to some the moral price being paid to "commodify" life is too high. These concerns, shared by many on both the left and the right, are central to the ways that biopolitics resists the traditional ideological spectrum.

CHAPTER FIVE

VALUING HUMANITY

BIOPOLITICS IS THE NONVIOLENT STRUGGLE FOR CONTROL OVER actual and imagined achievements of the new biology and the new world it symbolizes. Whether that world is seen as better or worse than our own, the very idea of such power stimulates deep reservations on both the left and the right about the implications of the post-Enlightenment, scientific worldview. In many ways, the left-right axis that originated in a few legislative councils in post–Revolutionary France has proven remarkably adaptable, even as the defining principles of those on each end of the spectrum have changed. What has endured is a relatively greater identification on the left with improvements and protections for the less privileged and a greater support on the right for the traditional order.

But when the full range of biopolitical issues are considered along with the motivating concerns behind them, the picture turns out to be far more complicated than is captured by the left-right spectrum. Consider, for example, those who might broadly be identified as bioconservatives: they worry that the life sciences will modify human abilities in comparison with some natural norm. But there are bioconservatives on the right and on the left. Bioconservatives on the right emphasize that a loss of traditional values could result, especially the dignity that should be ascribed to all persons. Bioconservatives on the left focus on the possibility that social inequalities and ecological problems could be grievously aggravated by biotechnological innovations.

A particular pair of linked themes cuts across the left-right spectrum and at least superficially unifies the bioconservatives: commodification and alienation. These themes are especially important to bioconservatives, they are ubiquitous in modern philosophy and social theory. It is almost impossible to find a modern social science analysis of the body or of the way that human beings relate to human bodies that does not presuppose the idea of commodification or the idea that markets are capable of turning the body and its parts into items that may be bought and sold.

Commodification as a concept of political economy is traceable to Marxism. The anthropologist Lesley Sharp, for example, draws on Marx and other leftist theorists to assess biotechnologies "whose application in clinical and other related scientific arenas marks a paradigmatic shift in anthropological understandings of the commodified, fragmented body." Examples include "reproductive technologies; organ transplantation; cosmetic and transsexual surgeries; genetics and immunology; and, finally, the category of the cyborg." Similarly, the bioethicist Donna Dickenson has argued that,

> One effect of late capitalism—the commodification of practically everything—is to knock down the Chinese walls between the natural and productive realms, to use a Marxist framework. Women's labour in egg extraction and "surrogate" motherhood might then be seen as what it is, labour which produces something of value. But this does not necessarily mean that women will benefit from the commodification of practically everything, in either North or South. In the newly developing biotechnologies involving stem cells, the reverse is more likely, particularly given the shortage in the North of the egg donors who will be increasingly necessary to therapeutic cloning.

In this tradition the commodification of the body and its parts may also contribute to social alienation. For Marx, alienation is the distancing of man from his labor, from his product, from other men, and finally even from himself: "the product of the laborer stands in no immediate relationship to his need and to his status but is rather determined in both directions through social combinations alien to the laborer." The worker's product is externalized as money or a good to be sold, reduced to a mere thing rather than an expression of individual creativity and desire. Alienation is the direct result of structural problems in capitalism, in which all products of human effort and human beings themselves are assessed in terms of their market value. Because the worker's body is the instrument of production—of the creation of new wealth or capital—under capitalism, the worker is ultimately alienated even from his own body.

Neoconservative Roots

Cultural conservatives who write about biology have thoroughly integrated commodification and alienation as core themes, even though these have Marxist origins and are normally identified as conventions of academic liberalism. To appreciate how this has happened, one needs to understand the origins of American neoconservatism, or the "new right," which is the product of a generation of disillusioned Marxists. As members of the "old left," they accepted the Marx of *Das Kapital*, who argued that economic forces are the substructure that dictates the superstructure of culture: "[T]he mode of production of material life conditions the social, political and intellectual life process in general. It is not the consciousness of men that determines their being, but, on the contrary, their *social being that determines their consciousness.*" For Marx, economic materialism determines moral values.

But by the 1950s, many intellectuals, including especially many talented Jewish thinkers, were driven from their early Marxist materialism and sympathy for Soviet Communism by the crimes of Stalin and Mao. In the late 1930s, the Communists were the most vigorous opponents of Nazism, so Communism was a natural refuge for Jews and other humanist activists. However, after the Hitler-Stalin pact, disillusionment ran high. Those who held on to their Marxism-Leninism through that crisis of confidence were gripped by a nearly religious fervor. This was classically described in *Witness*, the memoir by former Communist Whittaker Chambers, about his conversion from Communism to Christianity and his testimony against Soviet agents in the United States.

Yet the gradual revelations of Stalin's crimes, which led to the deaths of millions of his own people, stimulated a conservative reaction. Convinced that the main lesson of their lost faith was the nonnegotiable importance of morality over and against economics, and freedom over state-enforced "equality," the new conservative thinkers in effect turned Marxian analysis on its head. They came to argue that it was the so-called superstructure of moral values that was and should be the true driver of history, even of economic arrangements. Neoconservatives believe these moral values are grounded in immutable human nature, including the need for personal incentives, which the failures of Communism had shown could not be changed by social policies or government fiat. "What is absolutely clear," according to Irving Kristol, who cofounded the important conservative journal *Commentary*, "is that socialism turned out to be utterly unsuited to the nature of modern man." Francis Fukuyama, a member of the next generation who was at one time more identified with the neoconservatives than he is now, seconds this point: "Socialism foundered . . . on the shoals of a human nature that wouldn't let utopian planners do as they wished." Instead, neoconservatives admire the success of America's World War II generation, which they saw as proof of the indispensable moral superiority of America's open society grounded in a static conception of human nature and morality.

The neoconservatives' inversion of the Marx of *Das Kapital* continues to owe much to Marx's early writings. The first generation of neoconservative thinkers identified themselves more with the concerns of the "young Marx" of the *Economic and Philosophical Manuscripts of 1844* than the "old Marx" who wrote under the influence of Ricardo and the other classical economists. This young Marx, rediscovered by European left-wing students during the late 1960s, was much more of a philosopher and romantic idealist than the materialist economist of *Kapital*. In his early writings, Marx worried about capitalism's corrosive effects on human nature and human relations, asserting that "money externalizes the mediation between people, thus alienating it from them." He was concerned with the problems of alienation and commodification that he associated with capitalism. These were concepts that would prove influential, if in a somewhat altered form, in twentieth-century radical social criticism, including in the allegations by bioconservatives with regard to biotechnology.

We have become familiar with the foreign affairs idealism of the neoconservative thinkers who came to power in the George W. Bush administration. It was characterized by a confidence in the need for America to use its hard power to advance its democratic moral vision and a universal right of humankind that knows no ethnic or national boundaries. Claims that American forces would be "greeted as liberators" in Iraq reflect the neoconservative belief in fundamental human nature, that once people were freed from tyrannical government, they would adopt Western (and perhaps especially American) notions of government and civic virtue. Again, the notion of a fixed human nature can best be understood as a response to Marxian notions of human malleability before the material forces of history.

Rethinking Biotechnology

Doubts about the Bush administration's national security policy focused the world's attention on the nature of this moral idealism in international affairs. Less attention has been paid to neoconservatives' critique of the technologies that may issue from the new biology. This bioconservatism of the right is also founded on a secular moral idealism that esteems traditional values, but the result is a far more pessimistic concern with a distant dystopian future than is found in the fairly optimistic neoconservative philosophy of foreign affairs. Neoconservatives exhibit a deep ambivalence toward scientific innovation and sometimes even the implications of science itself for human well-being. They fear that technology commodifies and alienates man from himself and worry that the technological outlook obscures more important values. Adam Wolfson captures cultural conservative reservations about biotechnology:

> First, the continued development of biotechnology in certain directions will require the violation of truly basic moral strictures. Second, biotechnology will initiate a revolution in how we think about family, parenthood, the relation between the generations, work and achievement, and many other areas of human life. And third, biotechnology could bring about a fundamental rupture in human history, leading us into a "posthuman" age.

The violent language ("violation," "revolution," "rupture") is instructive. Neoconservatives worry about the changes wrought by biotechnology because they doubt the potential for true progress and fear that civilization is a fragile achievement. Carrying their own scars from failed attempts to change the world, neoconservatives like Gertrude Himmelfarb argue that "progress is not always lovely . . . [it] is unpredictable and undependable." They believe that supporters of biotechnology are making the same mistakes as Communists did fifty years ago in thinking they can create a qualitatively superior society. William Kristol, an influential conservative policy intellectual and editor of the *Weekly Standard* (and Irving Kristol's son), and Eric Cohen, former editor of *The New Atlantis*, explicitly draw the comparison to Communism, arguing that supporters of biotechnology exhibit "altogether an odd mixture of the hubris of the medical researcher seeking to lead his fellow men beyond nature, and the sentimentality of the post-Communist romantic, who seeks in genetic science man's new hope for building a kind, just and liberated heaven on earth."

But for neoconservatives, this hope is entirely misplaced, as the Soviet "experiment" should have taught us. Progress is difficult because people are too imperfect. The failures of socialism, and especially the catastrophic consequences of Communism, are seen as proof that people could not live together fairly and equally without moral guidance and market incentives that promote personal responsibility. The idealism of the early socialists led down the path of economic and moral ruin. Capitalism best reflects human nature but must be accompanied by a system that inculcates civic virtues, fostering what Adam Smith called the moral sentiments. The early neoconservatives were also deeply influenced by the tenets of psychoanalysis. As Irving Kristol laments in a passage reminiscent of Freud's *Civilization and Its Discontents*, "there is wealth enough for people to live fully and contented lives in socialist equality and fraternity—if only people wanted to. They do not. What they want is—more."

A striking lack of confidence in the human ability to manage the Promethean power of science characterizes neoconservative writings. For them, the new biology is a moral minefield because people are inherently imperfect, immoral, and weak. Unlike traditional American conservatives, many neoconservatives argue that even capitalism, generally associated with progress and individual ability, is grounded in human weakness. Cohen writes that "the moral defense of capitalism once rested firmly on a belief in the limited wisdom and virtue of human beings." Neoconservatives believe we must acknowledge these human weaknesses with biotechnology as well and therefore limit the ability of people to use biotechnology in the service of change.

This pessimism about people's ability to manage change is often focused on a dystopic, even apocalyptic vision. In inverting Marxism, neoconservatives depreciate confidence in a better future and in human potential. Instead, they use Marxist conceptions to analyze a society they see as in imminent danger of decline as technology impinges on long-standing human values. Often they share with Freud a strong Victorian streak in their attitudes about the future of human relations and the society it will produce. Perhaps Leon Kass captured their mood best when discussing the prospective lives of his beloved grandchildren: "I hope they will find pockets where they can enjoy what modernity has to offer without becoming its slave. But I wouldn't trade my life for theirs."

Neoconservatives are also concerned with man becoming distanced from his labor, from others, and from himself. But where Marx saw the influence of capitalism, neoconservatives argue that technology, not the economic structures that dictate its use, causes alienation. Thus, in vitro fertilization and other assisted reproductive technologies are criticized for distancing people from the procreative act and the recognition of their child as a fellow human being; they make us "come to think of children as mere products of our will." The theme is made explicit in Cohen's citing of Paul Ramsey: "To soar so high above an eminently human parenthood is inevitably to fall far below—into a vast technological alienation of man."

Writing in *The New Atlantis,* Paul Cella makes an even more expansive claim of the dangers of technological alienation. In a capitalist society, prosperity born of technology has a way of "spoiling" us, limiting our understanding of the difficulty and sacrifice inherent in work. "[The danger lies] in a diminution or impoverishment of the human things behind technology. . . . Modern men armed with the most sophisticated techniques and technologies, will lack any real understanding of the dignity of human work. . . . Part of this alienation lies in our remoteness from the causes of things, on the one hand, and from their consequences on the other." For Cella, the dignity of human work is not destroyed by "social combinations alien to the worker," but by the ease and remoteness of work due to technology.

Cella suggests these negative effects of technology are pervasive, even affecting the doctor-patient relationship.

> A certain alienation lingers in both parties as the human touch diminishes. The doctor's vocation is transformed from personal interaction and care to impersonal liaison between the enormous technological institutions of modern medicine and the patient. The doctor discovers one day that he is less a healer than a bureaucratic middleman.

Cella's words are remarkably similar to those of Marx and Engels in *The Communist Manifesto*, in which they decry a pervasive alienation that even affects the intellectual classes: "The bourgeoisie has stripped of its halo every occupation hitherto honoured and looked up to with reverent awe. It has converted the physician, the lawyer, the priest, the poet, the man of science, into its paid wage labourers." Beginning in England in the early nineteenth century and continuing through Marx, Nietzsche, Durkheim, and Freud to Bob Dylan and the Sex Pistols, alienation has been a common theme among observers of industrializing societies as the relationship between work and its product is subordinated to machine efficiency. Cella's HMO staff doctor is as much at risk as the assembly line worker.

Problems with Capitalism

Perhaps a more surprising agreement between Cella and Marx is the belief that monopoly capitalism contributes to alienation. Granting "the importance of largeness in the free enterprise system," Cella nonetheless "cannot see how liberty is best preserved in the implacable swallowing up of small, autonomous firms into vast bureaucratic corporations." This is a point he shares with early twentieth-century American progressives, including Justice Louis Brandeis, who railed against "the curse of bignesss" in corporate America. Yet Cella does not focus on the structural problems in the capitalist system that gives rise to these monopolies. Fully inverting Marx's conclusion, he argues that private property is the key to combating alienation because it is "something purchased on the credit of honest labor, is a near constant reminder of the ultimate humanness of all our achievements." Ownership gives life meaning.

The focus on technology rather than the economic and political conditions in which that technology is used is a typical problem in neoconservative writing. Such an analysis would be awkward for neoconservatives, who see economics as just a function of morality. Cella's own comments on the matter are telling: "It is precisely backwards to let economics dictate our principles—for economics is a tool." It is human nature and morality that dictate the shape of one's life, not impersonal economics.

Consider this condemnation of technology-induced alienation from *Beyond Therapy*, a report in 2003 by the President's Council on Bioethics during President George W. Bush's first term. This passage reads like a radical Marxist screed.

> What matters is that we produce . . . in a human way as human beings, not simply as inputs who produce outputs. . . . If all that matters is getting more out of [workers]—or more out of ourselves, by any means possible—then improving performance by biotechnical intervention makes perfect sense. . . . What matters is that we do our work and treat our fellow workers in ways that honor all of us as agents and makers, demanding our own best possible performance, to be sure, but our best performance as human beings, not animals or machines.

Despite these strong words, the report fails to consider the structural causes of this alienation. As Brandon Keim wrote in his review of *Beyond Therapy*, "that the political and economic environment in which biotechnologies will emerge deserves to be critiqued as thoroughly as the biotechnologies themselves seems only logical, but the recommendation is never made." Neoconservatives are obsessed with the depersonalizing consequences of alienation, but their focus on proximate causes instead of underlying social and economic dynamics renders their analysis incomplete. The lack of attention to the human consequences of a market-driven economy is understandable; it can be explained by the tension within the neoconservative critique of biotechnology.

Neoconservative bioethicists also use the Marxist notion of commodification in their attack on biotechnology. Marx believed that capitalism inevitably turns all human relationships, even our relation to our own bodies and to those of others, into monetary ones. The *Communist Manifesto* argues that capitalism has replaced the earlier idea of "natural superiority," which kept people in their place, with the brutal exploitation of cash payment. No sector of human life is immune, even the "holy family" of the Protestant ethic: "The bourgeoisie has torn away from the family its sentimental veil, and has reduced the family relation into a mere money relation." According to Marx and often for the neoconservatives as well, in capitalism every person and every social institution is valued only insofar as it is a source of further capital.

Neoconservatives express equally dire fears about the effects of commodification on human society. The root cause for these concerns, however, is once again new biotechnologies, particularly reproductive techniques. Cohen warns, "the most striking dimension of the modern economy is the commerce of the body, including an impressive array of new biotechnologies and biological procedures that promise to improve, control, or manipulate our native biology." Kass argues that "[with cloning] the commodification of human life will be unstoppable." Conservatives fear that an increased power over reproduction, not just by laboratory scientists but by individuals too, will make "human reproduction take on the semblance of manufacture, and parents would come to think of themselves as 'smart shoppers,'" the ultimate form of commodification.

Yet while the neoconservative and radical left concerns about biotechnology are at first blush identical, neoconservatives are reluctant to address underlying socioeconomic forces. Neoconservatives focus on the technology, not the context in which it is used. Consider, for example, Eric Cohen's discussion of "the new commerce of the body." He addresses commodification by examining paid egg "donation," the market in medications for erectile dysfunction, and a future in which human embryos are "valuable medical commodities—harvested routinely as a source [of] stem cells. . . . And no doubt such embryos will trade in the market like any other commodity—perhaps even on the 'commodities exchange.'" One would think the blame for these problems lies partly with an economic system that allows everything to be priced and sold. Rather than blaming capialism, Cohen argues that this rampant commodification is the result of a 1960s countercultural and biotechnological shared "belief that human limits should be overcome, taboos are anathema, and human shame is an illusion." Capitalism allows things to be done that should not be done, but humility keeps such problems in check. For Cohen, as for neoconservative thinkers generally, the true driver of the problem is morality, not economics.

And the proximate villain is cultural permissiveness, not capitalism.

Finally, though, neoconservatives are in an awkward position with respect to capitalism. On one hand they deeply admire capitalism for the way it cultivates individual initiative and encourages self-discipline. They detest the moral consequences of socialism and loathe the prospect of an overt split with more market-oriented traditional conservatives. On the other hand, remembering the Great Depression, they reject unbridled capitalism and keep their distance from libertarians. Concerning science, they also harbor reservations about the way that capitalism stimulates and supports biotechnologies that debase human nature. The result is a fundamental ambivalence about capitalism. According to Cella, "prosperity paradoxically enervates the human virtues that gave rise to it in the first place." "Of course," Cohen reminds us (and perhaps himself), "most biotechnology is admirable; it is a continuation of bourgeois progress as we have long known it. . . . At the same time, however, we must face up to the fact that modern commerce is often a moral problem, the capitalism of the body most especially."

The German Philosophers Strike Back

Biopolitical conservatism has both Anglo-American and German roots. Bacon's *New Atlantis* is often credited as the first utopian vision of the modern idea of progress framed in terms of advancing science and technology. In Bacon's tract, the governor—who, it becomes clear, is not a governor in the modern political sense—explains that this remarkable society called Bensalem is "dedicated to the study of the works and creatures of God. . . . The end of our foundation is the knowledge of causes, and secret motions of things; and the enlarging of the bounds of human empire, to the effecting of all things possible." It is truly a society governed by knowledge and wisdom, the finest flowers of the essence of humanity. Following a recitation of their remarkable innovations, the governor mentions that "we have consultations, which of the inventions and experiences which we have discovered shall be published, and which not; and take all an oath of secrecy for the concealing of those which we think fit to keep secret; though some of those we do reveal sometime to the State, and some not." There are insights too powerful to share with the political system; the scientists determine the mysteries that may be disclosed.

The Enlightenment utopias, especially Bacon's, illustrate these philosophers' identification with and revival of themes in classical Greek thought. Plato structured his *Republic* so that only those with knowledge of the good would govern. They constitute a natural aristocracy to be entrusted with important truths: truths that would anyhow be only incompletely apprehended by lesser souls. By contrast, "democratic" rulers might stumble on some dangerous truths—and not because of some inherent virtue, but rather by accident. Although the imperfections and cacophonies of democracies tend to make them destined for corruption and thus short-lived, these inadequate governors might be around long enough to do great harm because they will be under the influence of the mob that spawned them. The true rulers of Bensalem are the scientists, who are sometimes obliged to take it upon themselves to withhold more sensitive and powerful knowledge from those in charge of the state.

Whatever the source of their discomfort, whether political insurrection or the growth of knowledge, Bacon and the eighteenth-century British statesman Edmund Burke shared a preference for stability grounded in intellectual confidence about the order of things. Yet the experimental method and the liberal equality it fostered aroused demands for change. Because it vindicated demonstration over authority as the source of knowledge, the Enlightenment produced subversive ways of thinking, but it retained, elaborated, and vastly improved upon the ancients' appreciation of deduction and mathematical order. This latter tendency characterizes much of Western philosophy, what John Dewey would later call the quest for certainty. The deep fear of disorder, both intellectual and societal, is a familiar theme in post-Enlightenment philosophy, specifically among social conservatives. In this sense progressive thought is always suspect, especially to the extent that it is associated with science.

During the same period, from about the late eighteenth through the late nineteenth centuries, the German philosophical tradition developed a somewhat different approach. The Germans appreciated that the implications of observation, experimentation, and demonstration had already vastly changed both daily life and the human experience of the universe. In some respects, they were able to observe these effects in a more detached way than their British counterparts, as the decentralized German states had not directly participated in their own Enlightenment.

But German philosophers were well aware of the thinking to their west. Immanuel Kant said he was awakened from his "dogmatic slumber" by David Hume's observation that we never perceive alleged physical laws like causation, only a succession of discrete events. Kant saw the threat that this radically empirical approach posed for spiritual life and religious sensibilities. How could God be protected from empiricism? Kant's famous response in his *Critique of Pure Reason* was to establish an architecture of the world of phenomena or appearances, one that was ordered by certain a priori categories like causation. Of course, since the world of appearances cannot have come from nothing, Kant needed to postulate a realm of things-in-themselves, a noumenal world that by definition can never be known directly. God can then be inferred as a regulatory idea necessary to make sense of that which lies behind the realm of appearances. Thus, belief in that which is beyond our experience can be saved from a cold and dry empiricism. Hegel later generalized individual phenomenal experience to a collective consciousness that acts as a driver of history. The logic of both the Kantian and Hegelian philosophies, however different in their details, leads them to a positive construal of the prospects for human experience, even after the imposition of scientifically derived categories.

Kant inspired two important successors in the German philosophical tradition, Schopenhauer and Nietzsche. While they responded to the power of post-Enlightenment science in somewhat different ways, both were far more critical and pessimistic about the prospects of rescuing faith from the scientific worldview. In *The World as Will and Representation,* Schopenhauer contended that Kant's realm of things-in-themselves is what he called "the will," a metaphysical collective consciousness of base and irrational human desires that ultimately and unavoidably cause human life to be filled with suffering. Operating under Schopenhauer's influence since he was a young man, Nietzsche found it impossible to be honest about the implications of the post-Copernican worldview without acknowledging its catastrophic implications for human values. With the destruction of a comprehensive worldview that puts humanity at the center of creation, the entire story of Genesis and all the inferences that can be drawn from it, including our spiritual and moral norms, must be given up. Only a "super-man" who transcends this sorry outcome of the human story can retrieve a realm of values, for only one who transcends history and who can defy gravity (a literal and figurative weight of post-Enlightenment man) has the courage and spirit to legislate his own values. Meanwhile, the only way to avoid nihilism is to embrace the pre-rational musical drama of Greek tragedy.

Modern conservative responses to science have been deeply influenced by Nietzsche, who dismissed as "Socratism" the idea that science and technology can solve all human problems. What if Nietzsche was right about the moral consequences of looking honestly at the loss of human centrality in the universe? (Nietzsche himself was deeply impressed by the achievements of modern science and technology, taking a position on the theory of knowledge resembling that of the American pragmatists.) For neoconservatives, the implications of a widespread embrace of such a philosophy warrant the reassertion of traditional values. Here is where Burkean conservatism incorporates the influence of German existentialism into neoconservative thought. The way out of the moral abyss presented by scientific thinking is not by embracing the super-man like the transhumanists did (more on them later), but through cultural traditionalism.

Against Technology

Neoconservative worries that alienation and commodification are caused by technology stem from a worldview that mixes Marxism with the philosophy of Martin Heidegger, who noted that there are many different ways of finding meaning in the world, including art and religion as well as science. Reminiscent of Kant, Schopenhauer, and Nietzsche, Heidegger argued that none of these ways of finding meaning reveal the entire truth. Quite the contrary: by revealing some aspects of the world, these human activities conceal others. The technological worldview is particularly problematic for Heidegger, as it was for Nietzsche, because manipulating the world obscures the fact that doing so is only one of many ways of viewing it. Science and technology seek to explain everything, leaving no room for other means of explanation Heidegger wrote, "Technology threatens revealing, threatens it with the possibility that all revealing will be consumed in ordering," according to the dictates of science. Man comes to believe science is the only means of expressing truth, thereby excluding other methods.

Heidegger believed technology and science amplify the problem in that they approach the world in an artificial way. Objects in the world are viewed as things to be used for specific purposes; all human use of nature, no matter how seemingly benign, treats the world as a mere "standing-reserve" for use at human discretion. As David Tabachnick explains, "Rather than accepting that plants, human beings, cultures, or even war have a given nature or essence, technology treats all things as 'stuff' to be manipulated."

Heidegger saw the technological worldview as one in which nature is to be mastered and used for human ends. He believed this worldview would inevitably lead to further attempts at mastery, culminating in violations of human dignity. In perhaps his most explicit condemnation of technology, Heidegger wrote that "agriculture is now a motorized food industry—in essence the same thing as the manufacture of corpses in the gas chambers and extermination camps, the same thing as the blockading and starvation of nations, the same thing as the manufacture of hydrogen bombs." Comparing modern agriculture to gas chambers and death camps takes to a wild metaphysical extreme the Luddites' destruction of textile machines that threatened their livelihood in the early nineteenth century.

The Heideggerian critique of technology as a dangerous worldview is central to neoconservatives' critique of biotechnology. Kass, who has long expressed reservations about fertility treatment, organ transplants, and of course human cloning was deeply influenced by Hans Jonas, who was in turn influenced by Heidegger. Kass's writings are characterized by worries about the threats technological interventions pose for human dignity. But in fact, the concern is the still deeper notion that technology tends to alienate us from the very core of our being. Similar to Cella's concerns about alienation, Kass argues that "the so-called empirical science of nature is, as actually experienced, the highly contrived encounter with apparatus, measuring devices, pointer readings, and numbers; nature in its ordinary course is virtually never encountered directly." Worse yet for Kass, "modern science rejects, as meaningless or useless, questions that cannot be answered by the application of method." Science excludes the very moral questions that neoconservatives view as the driver of history.

Kass is not the only leading neoconservative critic of biotechnology whose thought is in some substantial measure Heideggerian. Francis Fukuyama uses a Heideggerian epigram for the first chapter of his book, *Our Posthuman Future*:

> The threat to man does not come in the first instance from the potentially lethal machines and apparatus of technology. The actual threat has always afflicted man in his essence. The rule of enframing (Gestell) threatens man with the possibility that it could be denied to him to enter into a more original revealing and hence to experience the call of a more primal truth.

Eric Cohen laments that "one of the great shortcomings of modern society is that seeking remedies (or technology therapy) becomes our overriding aim, crowding out the search for wisdom, love, excellence, and holiness that is central to living a full human life." Neoconservatives are not merely concerned with particular technological innovations, but the nature of technological advance *in toto*.

This tension between their antitechnological stance and an implicit acknowledgement of the undeniable benefits of technology permeate neoconservative writings about bioethics. Cohen attempts to refocus the problem on the excesses of technology: "Our mission and our moment is [sic] inseparable from facing up to our technological condition: savoring it, defending it, and improving it, but also coping with it, transcending it, and reining it in." Yet this position conflicts with his view that it is the technological attitude as a whole, not simply certain aspects of technology, that is the problem. Kass seems to more fully accept the implications of his underlying philosophy when he writes that "there are no more deep unfulfilled human wishes for which technology of the future is going to provide the answer . . . we already have more than what we need to live well."

Neoconservatives worry about the dangers of technology yet provide no criteria for distinguishing the destructive technologies from those that do not threaten human dignity. The technological attitude does not end with stem cell research, but is also present with regard to automobiles, vaccines, the World Wide Web, and virtually the entirety of modern life. Such a generic antitechnological posture, while at first seductive, is not only extreme, but offers no guidance for practice. For neoconservatives, the potent social and economic forces that capitalism unleashes are to be constrained by certain traditional moral standards. As one social scientist has noted of the neoconservatives, they aim to "remoralize" our benighted postindustrial and dangerously posthuman society through the application of Judeo-Christian ethics.

Yet one senses a deep sadness in much neoconservative commentary about science. It is as though they are aware that, at the end of the day, economic pressures and sheer hubris encourage people to act contrary to what they regard as humanizing values. They express the morose sentiment that what lies ahead seems unavoidable. In the final analysis, these bioethical neoconservatives are the descendents of Schopenhauer's deeply pessimistic view of human life as one of inevitable suffering rather than the problem-solving and practical sensibility of American pragmatists like William James and John Dewey. None of man's hubristic efforts can avoid the tragic reality behind the temporarily distracting appearances. If he could stop all embryo research he would probably do so, Cohen writes, conceding that he cannot. "We can, however, try to stop the worst horrors—and worst temptations—before they arrive. . . ."

Taking Precautions

Grim as it is, perhaps "just say no" to technology isn't the neoconservative message. A more charitable interpretation of neoconservative reservations about technology could take them as supporting a version of the precautionary principle. Mainly identified with European environmental activists, the precautionary principle holds that if an action or policy could cause serious and irreversible harm to the public, the burden of proof that the safety margins are acceptable falls on those who advocate that action or policy. Often the principle is interpreted as requiring that the advocates be able to present some scientific consensus that no harm will follow. Proving a negative is at best not an easy task.

During the 1980s the prudential principle became a rallying point for European critics of international corporate giants' exploitation of the underdeveloped world after the Bhopal disaster. An accidental release of toxins from a Union Carbide pesticide plant in India resulted in thousands of deaths and hundreds of thousands of exposures. In *Risk Society*, German sociologist Ulrich Beck took Bhopal as the exemplary case of his theory that a characteristic of modernity is that it exports risk; in this case, the direction was from a U.S. multinational to a poor country on the other side of the planet. According to Anthony Giddens, the result of such a system of risk production is a society that is increasingly preoccupied with threats to safety, a "risk society." Though stimulated by examples of environmental catastrophe rather than the power of the new biology, Beck's notion of a risk society is closely related to the idea of biopolitics. The common thread is competition among various forces that face public pressure to take control over the emerging science and its applications or face the loss of public confidence and legitimacy.

As far as I know, none of the prominent neoconservative writers about bioethics have explicitly associated themselves with the precautionary principle. To do so would put them in a somewhat awkward tactical position. Unlike a growing number of Christian conservatives who worry about climate change, the neoconservatives are deeply skeptical of the environmental movement, often because of its association with leftist European intellectuals. Yuval Levin has theorized that Europeans use allegations about climate change as an excuse to criticize American foreign policy. Even though it makes for strange bedfellows, a radical precautionism is at least superficially in step with both the neoconservative and green progressive critiques of biotechnology.

But the ecologically minded progressives often called "greens" turn out to be less radical in their precautionary sensibilities than the neoconservatives. Theirs is a cautionary tale about consequences, not a grim metaphysical conclusion under the influence of German existentialism. The greens' critique of proscience assumptions is at its sharpest when confronting what they regard an unwillingness to consider the potential bad outcomes of innovation. These include the effects on those who are vulnerable and unable to participate in the benefits of technological change. Rather than viewing the precautionary principle as a means of opposing new technologies generally, they see it as a way to promote public health and the environment over corporate profiteering, noting that some version of the principle has been integrated into virtually all international environmental agreements. In the words of Marcie Darnovsky, "[e]nvironmentalists and other progressives who support a precautionary approach readily support beneficial applications of new technologies, and acknowledge that we already enjoy many of them." But she continues,

> [W]e must also be mindful of many reasons for concern, from the great civilizations that have collapsed because of misuse of contemporaneous technologies, to the twentieth-century genocides that depended on harnessing the scientific and medical establishments of that era, to the severe environmental degradation that plagues many parts of the world, to the prospect of globally catastrophic climate change.

Similar worries from the left are expressed by the ETC Group, an international advocacy organization that describes itself as "dedicated to the conservation and sustainable advancement of cultural and ecological diversity and human rights." Besides examining familiar issues like brain manipulation through neuroscience and the possibility that biotechnology will produce new biological weapons, ETC looks to the implications of biotechnology for indigenous agricultural economies and cultural practices as well as environmental consequences. Of particular concern to ETC is that powerful agricultural interests and industry scientists could combine to create products like "suicide seeds," genetically modified to prevent farmers from reusing seed stock, or products of synthetic biology that could threaten the livelihoods of subsistence farmers in the developing world and create massive new burdens for local water supplies. Like other such green progressives, ETC sees itself as providing a cautionary counterweight to the massive financial resources and political clout of industry and the globalized system of scientists and engineers whose services it can buy.

Darnovsky contrasts this cautious attitude with a "mythic" view of science that, she argues, has been adopted by some on the left. According to Darnovsky, some progressives and liberals took an aggressively proscience position partly as a reaction to the Bush administration, a position that identifies advances in science and technology with progress itself. Attempting to correct what they regard as excessive scientism on the left, green progressives often sound very much like cultural conservatives, who also decry threats to human dignity and rampant commodification. In this sense, both groups may be regarded as "bioconservative."

But this superficial resemblance does not tell the full story. In response to the emphasis on individual autonomy in academic bioethics, which is grounded in hospital ethics and the patients' rights movement, Darnovsky urges a greater emphasis among progressives on social justice, the public interest, and the common good. Unlike neoconservatives, the bioethics greens have a far more concrete metric of human dignity that is grounded in the fair distribution of social goods and a rejection of class- and race-based privileges. Their critique of capitalism is vastly more explicit than that of the neoconservatives, which has made them a nice target for proscience libertarians who see them as betraying too much sympathy for cultural conservative values.

Because they are more consequentialist in their reservations about technology than neoconservatives, green progressives should be more amenable to strategies that ameliorate the effects of biologically based innovations. Ultimately, the greens provide an important reminder of the human costs of the concentration of wealth and power that is often associated with science-based innovation. They are correct that bioprogressives would err in rejecting human dignity as an irreplaceable value, however amorphous. But they fail to acknowledge the larger role of science as a form of inquiry that has cleared the way for a modern conception of human rights.

All Too Human?

The greens' critique of procience assumptions is at its sharpest when confronting "technoprogressivism," a view defined by James J. Hughes as "the consistent application of progressive values to technology and enhancement." Once again, Francis Bacon's *New Atlantis* is a kind of lodestar, as the governors of his utopia believed that virtually all aspects of human life could be drastically improved by the application of science. Some technoprogressives are transhumanists, also called posthumanists. In their admiration for science and their notion that human beings need to transcend the narrow confines of humanity, perhaps by engineering the next phase of human evolution, this small circle of scientists and philosophers is indebted to both Bacon and Nietzsche. According to Nick Bostrom, "[t]ranshumanism holds that current human nature is improvable through the use of applied science and other rational methods, which may make it possible to increase human health-span, extend our intellectual and physical capacities, and give us increased control over our own mental states and moods." As parties in the struggle for biopower, the posthumanists seem to be a rather obscure and insignificant threat, but they have been vigorously targeted by bioconservatives of the left and the right. From the left, three scholars write that

> The new species, or "posthuman," will likely view the old "normal" humans as inferior, even savages, and fit for slavery or slaughter. The normals, on the other hand, may see the posthumans as a threat and if they can, may engage in a preemptive strike by killing the posthumans before they themselves are killed or enslaved by them. It is ultimately this predictable potential for genocide that makes species-altering experiments potential weapons of mass destruction, and makes the unaccountable genetic engineer a potential bioterrorist.

From the right, Francis Fukuyama calls transhumanism "a strange liberation movement" that wants "nothing less than to liberate the human race from its biological constraints." He writes, "The first victim of transhumanism might be equality. . . . If we start transforming ourselves into something superior, what rights will these enhanced creatures claim, and what rights will they possess when compared to those left behind?" As if that were not alarming enough, Fukuyama calls transhumanism "the world's most dangerous idea."

Since neither the probability nor the magnitude of the threat posed by transhumanism is at all clear, the anxiety it provokes on the part of such ideologically disparate observers must be attributed to the symbolism of the new biology and some transhumanists' willingness to unleash the forces it represents. "I say, bring on those genetic bulldozers and psychotropic shopping malls that help people to live healthier, smarter, and happier lives," says libertarian Ronald Bailey in his response to Fukuyama. As to the potential for inequalities, Bailey counts on the core Enlightenment value of tolerance for differences to persist. In any case, he writes, "Let's face it, plenty of unenhanced humans have been quite capable of believing that millions of their fellow unenhanced humans were inferiors who needed to be eradicated." Biotechnological species improvements couldn't take all the blame for man's inhumanity to man, and they might just ameliorate them.

Not all transhumanists are unrestrained biopolitical libertarians, but they do tend to place their bets on a qualitatively different form of human life in the not-too-distant future. Worrisome to bioconservatives, especially those on the right, is that Americans in their enthusiasm for technological progress will happily put aside their values as they descend a slippery slope. Whether whatever lies at the end of that slope is as grim and deracinated as the neoconservatives fear, they are right that the future appeal of radical technoprogressivism might well lie in America's complicated love affair with progress.

CHAPTER SIX

CROSING LINES

OUR CONTINUING BATTLES OVER THE LIMITS OF CUTTING-EDGE research and power over the beginning and end of life are modern examples of biopolitical debates that began well before the recent science of bodily tissues and cells. But experimental biology and its related technologies are increasingly affecting the way various forces contend for control over life's familiar margins and biology's traditional boundaries. Long before the new biopolitics took its place on the public stage, Leon Kass summarized the problem from the neoconservative bioethical standpoint that would later set the biopolitical agenda:

> Science essentially endangers society by endangering the supremacy of its ruling beliefs. . . . Science—however much it contributes to health, wealth and safety—is neither in spirit nor in manner friendly to the moral and civic education of human beings and citizens. Science fosters and encourages novelty; political society, governed by the rule of law, cannot do without stability. Science rejects all authority save the truth; and prefers skepticism to truth . . . the political community requires trust in, submission to, and even reverence for its ruling beliefs and practices.

The notion that science is an enemy of moral and civic education is puzzling. How then to account for the coincidence of the development of science with the growth of liberal democracy and the recognition of universal human rights since the eighteenth century? Nonetheless, Kass is surely right that science upends prejudices and obscures familiar conceptual benchmarks. As Charles Peirce appreciated, that is precisely the point of science. Science, he wrote, "does not consist so much in knowing, not even in organized knowledge, as it does in diligent inquiry into truth for truth's sake, without any sort of axe to grind, nor for the sake of the delight of contemplating it, but from the impulse to penetrate into the reason of things. . . ."

Democratic institutions have a role in protecting threatening views of minorities, including minorities that are composed of scientific experts. The lawyer and bioethicist R. Alta Charo has argued provocatively that the American constitutional system, and particularly the First Amendment, should be seen as protecting basic science research as a form of free speech. Legal institutions may indeed be required when familiar and cherished worldviews are threatened. In particular, the challenge of obscured natural boundaries arises in a range of biopolitical issues like the long-standing public controversies about abortion and end-of-life decisions, the use of predictive testing for serious disease, the creation of lab animals with human tissues, and the new field of synthetic biology.

Abortion

I have often asked my students to consider this scenario: A fellow student announces at the beginning of class that, later that day, there will be a rally at the main campus plaza about genetically modified organisms (GMO) and agriculture. What would they predict would be the likely size of the crowd, regardless of whether the rally would be pro or anti-GMO? In general, my students agree attendance would be small. Now I ask them to imagine the same scenario taking place at virtually any major western European university. American college students are often surprised to learn that the crowd would be large and vocal, especially as a protest against GMO and corporate interests in food production.

Then I ask my students to substitute the topic of abortion for genetically modified organisms, again, regardless of side selected. And, again, I tell them to compare predictions for the likely size of the rally at an American and a European institution of higher learning. Abortion draws a lot of interest in the United States, but not so for European youth. Again, my students are often surprised that abortion is not the axis point of European biopolitics that it is here, but GMO is. And, in truth, I have a hard time explaining the many cultural and historical reasons for the differences. Part of the answer lies in western Europeans' more permissive attitude toward abortion, and that surely has to do with the greater secularity of western Europeans than Americans. On the other hand, for some European activists, the anti-GMO movement is a proxy for opposition to American corporate imperialism, especially in the form of companies like Monsanto. (There might also be a pinch of well justified Old World pride in their cuisine and its local, "natural" content.) But whatever the explanation, these examples illustrate how different American and European biopolitics are, that the ways that cultural suspicions about science or scientists manifest themselves depend on intersecting concerns that are hard to compare. In America, the question is to whom rights should be attached; in Europe, it is how corporate interests conflict with human solidarity and avoidance of risk.

Because of the importance of abortion in American biopolitics, it's important to gain some perspective on the issue to see how it might affect the future of biopoltical controversy. Since the U.S. Supreme Court's decision in *Roe v. Wade*, the question of the human embryo's moral status has cast a long shadow over politics in America. Even some Christian conservative bioethicists privately bemoan this turn of events, as other issues of concern to them, such as global justice, go begging for attention while human and financial resources are focused on the embryo debate. They recognize that at best the pro-life and pro-choice movements are locked in a mutually reinforcing struggle that tends to guarantee perpetual conflict. This is the bioethics version of the Israeli-Palestinian problem. As Leon Kass, who was the first chair of President Bush's bioethics council put it, the council came forth in the midst of "embryoville."

Although the abortion controversy will simmer on indefinitely at a formal political level—partly because it has been institutionalized in advocacy by both sides—its likely spillover effects on future biopower are unclear. Data suggests that Americans have settled into an often grudging acceptance of abortion on demand, though with various controls imposed, both official and unofficial. The official controls include state laws that put up some obstacles, especially for minors; the unofficial controls include the absence of abortion providers in some parts of the country. The pro-life movement has been especially successful at stigmatizing, and to some degree intimidating, medical professionals who might be prepared to perform the procedure. The U.S. abortion rate has in fact been decreasing—but it is not so much because of professional reticence as improved and more widely practiced birth control.

With some significant exceptions, academic bioethicists have not been among the most vigorous voices on abortion, either pro-life or pro-choice. After all, the academic's goal is to offer a new position or at least a novel argument, but the issue has been so politically polarized that there is little space to entertain or even conceive of an original idea. One of the few is Daniel Callahan's argument on behalf of an anti-abortion/pro-choice position. He believes abortion is morally wrong but he is not so confident in his belief that he is prepared to make it a matter of law, a position he issues as a challenge to the rest of us: are we so sure abortion should only be a matter of choice that we wholeheartedly endorse it without expecting some moral reflection? On the other hand, as police power is always a heavy-handed approach to moral enforcement, are we so sure that it is wrong in so many cases that we would translate that view into law? The fact that such a reasoned position can't even attract serious discussion illustrates the seemingly hopeless deadlock between the customary views.

In spite of the deep and persisting divide among many Americans on the embryo's moral status or on the circumstances under which abortion is justifiable, few observers believe that *Roe* will ever be fully reversed. Rather, survey data suggests that Americans are trending toward a wish for greater justification for particular abortions, reflected in state-level efforts to constrain and discourage abortion through the legal system. A growing number of pro-choice advocates also acknowledge that medicine's ability to provide images of the developing fetus in the womb, and even to perform prenatal surgery that in effect turns the fetus into a patient, has influenced the way people think about the fetus, including many who are pro-choice. But, again, the dominant result is restriction, not prohibition. This result suggests that most Americans are not "vitalists" who oppose abortion without regard for any other considerations. In particular, privacy is more highly valued than Americans' level of confidence about the moment when the embryo or fetus acquires rights—or, for that matter, about the "right to life" in general.

Yet even if abortion rights continue to become more restricted, it is not at all clear that the same will be true about research involving embryos, embryonic cells, materials derived from those cells, or even fetuses. The reason is the value Americans place on privacy and medical progress. After all, in spite of the restrictive legislation passed in some states in the forty years since *Roe* and the various controversies about federal funding for fertility research, the new biology has proceeded with astonishing momentum from one technical advance to another. For that to change, and so long as most people are comfortable that there is substantial moral space between abortion and medical research, much more draconian restrictions on the new biology would have to be introduced than has been the case so far. That comfort level is not always easy to achieve, as the embryonic stem cell debate illustrates.

Privacy, Power, and End-of-Life Decisions

The new biopolitics is about control over the tissues that make up bodies; the new biology will also force more difficult questions of control over bodies themselves, especially at the end of life. The high value Americans place on privacy at life's margins, and especially family privacy, was vividly illustrated by the Terri Schiavo case, which culminated in 2005, in which a young woman diagnosed as irreversibly unconscious was at the center of a legal battle between her ex-husband and her parents. Her husband, who was found by numerous courts to be her authorized health care decision maker, believed that she would not want to continue in a condition of total lack of awareness, as had been the case for years, so he asked for her feeding tube to be removed. Her parents objected. Finally, a Florida court found no basis for removing Michael Schiavo from his position as his wife's guardian.

The ensuing bizarre and dispiriting public drama captured the country's attention as some politicians came to believe that there was capital in taking an aggressive right-to-life position in the case. The Florida legislature passed a law that gave Governor Jeb Bush the authority to order that her feeding tube be reinserted, which he did. The Florida Supreme Court found the law unconstitutional. Then, just when it seemed matters couldn't get worse, in a macabre twist Republican members of the U.S. Congress subpoenaed both Michael Schiavo and his unconscious wife to testify at a congressional hearing. Several senators and congressmen with medical degrees, including Senate majority learder Bill first, claimed they could tell Terri Schiavo was misdiagnosed without having examined her. In medical ethics this is considered a gross violation, but apparently not for elected officials who happen to be doctors. Finally, in a compromise, Congress passed a law that President Bush signed transferring the case to the federal courts, where her parents' final legal efforts failed. Terri Schiavo died on March 31, 2005, after which an autopsy showed extensive brain damage, so that there was no physical basis for any level of consciousness or rehabilitation.

The line of legal cases about decision making for patients like Schiavo began in 1976 with Karen Ann Quinlan, a twenty-one-year-old woman who collapsed after coming home from a party, apparently having consumed alcohol and valium. Quinlan had also lost considerable weight due to a strict diet in the days before. Gradually, she lapsed into what is called a persistent vegetative state, neither aware of herself or her surroundings nor dead or comatose. After a few months her parents, having consulted with their priest, decided she would not recover and asked the hospital to remove her from her ventilator. The hospital refused, fearing that such an action would bring on a murder charge. The New Jersey Supreme Court found that her parents had the right to act on what were presumed to be her wishes as they understood them, and that the ventilator was "extraordinary means" and not morally required, as defined in a papal letter from 1957. To the surprise of specialists, Quinlan lived without the ventilator for another nine years.

The Quinlan decision affirmed that even those who are unconscious have the right to have their wishes respected. In this way, it seemed to locate decision making power in the patient or, if he or she is severely disabled like Quinlan, some close surrogate. One result was an explosion of interest in how those surrogate decisions should be made, including living wills and advance medical directives. But an unresolved question was whether artificial feeding, which was not at issue in *Quinlan*, was also covered by the same standards. Were food and fluids delivered through a tube more like other technical medical treatments, which could be optional, or were they more like giving someone who can't feed themselves a spoonful of food? A 1990 legal case settled that question when the family of Nancy Cruzan, a young woman who was in a persistent vegetative state as the result of an auto accident, was finally determined to have met the standard of evidence of her previous wishes that was required in Missouri. Unlike Quinlan, she was not on a ventilator, but her feeding tube was removed.

By the early twenty-first century, the Quinlan-Cruzan legal and ethical doctrine reaffirmed in Schiavo that during extended periods of incapacity, persons have the right to determine the course of their own treatment. Furthermore, in the absence of written advance directives, those they have chosen as their representatives are the most appropriate agents to speak for them. An important aspect of this doctrine that was somewhat obscured in the Schiavo controversy is that it is neutral with respect to whether treatment in these tragic cases should be continued or not, so that it protects those who would prefer aggressive interventions as well as those who would not. In the Schiavo controversy, conservative activists and elected officials overinterpreted the sympathy of most Americans for some degree of pro-life protections. They failed to appreciate the even greater support for family privacy. All this is not to deny that many Americans, including some on the left like the Rev. Jess Jackson, were discomfited by Michael schiavo's decision to withdraw artificial feeding, but neither did they want the political process to intervene.

That Americans have largely been comfortable with the Quinlan framework and saw no good reason to overturn it (or, indeed, discerned an acceptable alternative) was perhaps one reason that attempts to insert a political process into the Schiavo matter were widely rejected. The sense that this was a private family matter gone terribly wrong, and one that had been addressed by nearly two dozen courts, was surely also a factor that contributed to the negative reaction to congressional and presidential interference. The depressing media exercise that followed, including a cable news deathwatch, was a vivid demonstration that political exertions on behalf of deeply held but abstracted cultural moral views can be both clumsy and embarrassing.

The Next Terri Schiavo Case?

Neurologists who study cases of vegetative states like those of Schiavo, Quinlan, and Cruzan have come to identify them as *permanently* vegetative. Patients lack both the clinical signs of any underlying awareness and the physical substratum required for consciousness. Both permanent and persistent vegetative states are included as one of many "disorders of consciousness," along with coma, dementia, and other conditions. But there are instances of disorders of consciousness that are more difficult to diagnose as cases of the permanent vegetative state. These are patients who might be minimally conscious, retaining some form or level of consciousness that in the past we have not been able to detect. But that may be changing. Applications of new brain-scanning technologies seem to provide new ways to measure neurological activity. New ways to stimulate the brain through a combination of drugs and devices may also make it possible to provide a "lighten" of their consciousness, enabling them to have at least some awareness of themselves and their surroundings, or perhaps to improve whatever minimal level they had before. These patients may be *persistently,* but not permanently, vegetative, in the sense that they may be able to recover a modicum of awareness.

These new treatments will inevitably create a new front in the culture wars about the right to die. There can be little doubt that a test case involving a persistently vegetative individual will at some point emerge, probably one that involves disagreements among family members about whether more tests and interventions should be done. Such a case would be significant, as it would raise the question of whether such a patient is truly in the unrecoverable state the patient wished to avoid. In *Cruzan*, the Supreme Court permitted Missouri to require "clear and convincing evidence" that an individual now in a permanent vegetative state would wish to discontinue food and hydration. Were an individual in a persistent, rather than permanent, vegetative state, should the state be permitted—or even required—to impose an even stricter standard?

One approach would embrace new technology for both diagnosing and treating disorders of consciousness and argue that such technology has advanced to the point that no one should be given up for dead simply because they cannot feed themselves. On this view, artificial feeding should be continued in the face of uncertainty about the patient's true medical and ethical status—a new "precautionary principle" applied to those who might have a very small chance of some awareness. It's not hard to imagine that, like those who sought to intervene in the Schiavo matter, certain advocates might insist upon more and more tests and an endless round of interventions, with little distinction between their moral and political aims, thus exploiting a case for biopolitical advantage. Such a result could nonetheless undermine the Quinlan consensus.

But when decisions must be made about what someone who is gravely ill might have wanted, the symbolic power of these new technologies is greater than what they can actually accomplish. All medical information can do is inform the decision making process, not substitute for it. Deciding whether to continue treatment of any sort remains the right of the patient or the surrogate acting for the patient. The good news is that emerging technologies promise to provide more information, so that appropriate decision makers can feel more confident that they are acting according to the patient's wishes and best interests. These points will require repetition throughout media coverage of a continuing story that will strike many as more ethically ambiguous than the Schiavo case. As the dispiriting "death panels" incident demonstrates, one well-publicized controversial decision to limit end-of-life treatment, even though ethically, medically, and legally justified, could compromise decades of work to enhance patient self-determination. As we learned in the stem cell debate and the surprising legal process that followed, no question of biopower is ever settled.

(Bio)marking Time

From Quinlan to Cruzan to Schiavo, the death-and-dying controversies have centered on people who have already experienced tragic circumstances. But medical science is developing more and more reliable methods for helping us learn what our own risks for serious disease happen to be. In 2010, an international team announced that it had found a way to identify Alzheimer's disease (AD) by testing spinal fluid. Perhaps most striking was the observation that, in the dispassionate language of medical science, "The unexpected presence of the AD signature in more than one-third of cognitively normal subjects suggests that AD pathology is active and detectable earlier than has heretofore been envisioned." In other words, it can with confidence be predicted of people with no discernible symptoms that they are on the path to developing Alzheimer's.

Why would anyone choose to undergo a spinal tap, itself no walk in the park, for such a grim diagnosis? This and other predictive tests, such as one for Parkinson's disease, illustrate the appeal of biopower and the desire of many individuals to control it for their deeply personal purposes. Plans for assisted suicide might be one element of a decision, but in the event, few people choose to follow through. Rather, many who will choose the test will do so out of a need for a sense of control, the chance to plan sensibly for the time they have, the opportunity to put their affairs in order, and to refocus their lives on what is really important to them. They might also want to volunteer for the clinical trials that will try to target the underlying mechanisms of AD, both to contribute to medical science and to preserve a sense of hope. And some might feel that they would be more motivated to make the dietary and other lifestyle changes that seem to delay or lessen the symptoms.

In the case of the Alzheimer's biomarker, the apparent asymmetry between knowledge and control will provoke a debate about whether the test should be offered at all and if so, under what conditions. The procedure will have to be standardized for doctors' offices. Eventually brain scanning should replace the unpleasant spinal tap, making it much more accessible to those who want to know their fate. There is a close analogy to Huntington's disease, which is in some ways an even more difficult situation since those at risk often know their exact odds, and their reproductive decisions are tied up with the testing question. Some choose to be tested, and some don't. In light of the Huntington's experience, future candidates for the AD test might be required to undergo a psychiatric assessment. Counseling following a positive result should be part of routine practice. Besides these basic medical ethics considerations, there are important public policy issues. The National Institute on Aging estimates that as many as 5.1 million Americans might have AD. Insurers are going to have to decide about their policies for supporting testing and related expenses. Families as well as patients will need support. In the long run, though, effective medical interventions will flow from this new assessment technology.

Many medical ethicists argue that there is no good reason to have long-term predictive tests for conditions like Alzheimer's. After all, changes in lifestyle like a healthier diet and more intellectual activity are advisable for all of us, and, at least so far, there is no surefire drug that can prevent AD symptoms. If one has an affected relative, one generally knows the risks of such diseases of aging anyway. Philosophically this is a sensible position, but the plain fact is that images and biological tests are so appealingly concrete that they will always be in demand, even if they don't really provide useful new information or alternatives. They stimulate the demand and competition for biopower, especially if these are pictures of my own brain or tests of my own bodily fluids.

Enter the Minotaur

One longer-term result of new and more technically complicated biopolitical conflicts over end-of-life decisions could be renewed focus on attitudes toward the value and dignity of human life. The end-of-life controversies and predictive tests like those being developed for Alzheimer's certainly cause us to reflect on these ultimate questions about being human. So will other technologies, mainly emerging from the laboratory setting, that are primarily about the point a nonhuman animal becomes "too human." Chimera and hybrids are likely to be another bioethical issue of increasing public salience.

In Greek mythology the Chimera was a fire-breathing combination of a lion, a snake, and a goat. Now the word is used by scientists to describe creatures containing cells from two or more genetically distinct sources. Some human beings are chimera. For instance, when two embryos merge in utero, they yield one person with two cell types and, perhaps, two blood types. Once a woman has been pregnant, she herself is technically speaking a chimera, carrying cells from her fetus, apparently for the rest of her life. Chimera can also cross species boundaries, as when an adult human is surgically implanted with a heart valve from a pig.

Chimera of various kinds have also been used in important medical research for decades. In addition to heart valves, pigskin grafts are used to minimize scarring in severe burn cases, and cow arterial grafts are used by vascular surgeons as alternatives to synthetics during arterial reconstruction. Using chimera, geneticists are learning how a single gene operates in a complex system. Chimera are being used to build a foundation in stem cell regenerative science, the study of the ability of stem cells to hone to and replace damaged tissue. For instance, in models of neurological disorders, human-primate and human-rat brain chimera have been used to test the feasibility of future neural stem cell treatment of Parkinson's disease and stroke. In addition to their role as model organisms, human-animal chimera are a fundamental part of understanding the basic biology of cellular development and viral disease. Human-mouse bone marrow transplants have been performed since the 1980s and have been part of studies of AIDS and leukemia.

The sheer variety of examples should suggest that the word *chimera* is more exotic than it is descriptive. Originally connoting an illusion, these lab animals are quite real, and some will be regarded as disquieting. The very strangeness of the word imparts a distracting mystery to the animal models to which it applies, which are not combinations of body parts like those of a lion, a snake, and a goat. No doubt the scientist who first applied the term innocently aspired to align these valuable creatures with an aura of poetry, but the net effect is to sustain long-standing cultural anxieties about monsters. In the ancient world, the sighting of a chimera was quite literally a bad sign. Add the modern laboratory environment, itself a black box to most people, along with the widely held notion that scientists are either eccentric or too damn smart and ambitious or all of the above. Words like these can be setbacks for science in the public mind, and they cohere all too comfortably with the legacy of Dr. Frankenstein.

In 2005, I gave a talk about stem cells and chimera to a group of congressional staff. Afterwards I was approached by an experienced legislative aide to a senior senator known to be a good friend of medical research. He asked me if I could send him a list of "normal sounding" chimera. Perhaps his concern was partly motivated by a question that had been raised by the distinguished Stanford University scientist Irving Weissman a few months before: what ethical issues would be raised when doing experiments that involve putting human brain cells into mice? The purpose of these experiments is to gain a better understanding of diseases like brain cancer. Yet some would be discomfited by the possibility, however remote, that some human properties could arise in a mouse whose brain was composed mainly of human neural cells. Hence came the inquiry from the senator's staffer.

The buzz about chimeras got a boost when Weissman asked his colleague Hank Greely, a Stanford law professor and ethicist, whether it would be ethical to create a mouse with a brain made entirely of human cells in order to test new drugs. True to the practices of prominent universities the world over, a committee chaired by Greely was formed to examine the question. As Greely explained in a 2005 interview, he and his colleagues on the committee concluded that, "If everything in the brain looked like a mouse brain, we said the experiment can go forward. If, on the other hand, there were things in the brain that looked wrong, that looked like human structures, or that looked like very odd misshapen mouse structures, our advice was: Stop the experiment and talk about it. Our view was it's much better to try to decide these issues with some facts in hand, rather than to say in advance, 'always go forward; never go forward.'"

Fusion and Confusion

Chimera are often confused with hybrids, the products of fusing sperm and egg from two or more species. For example, a mule is a hybrid, as the sterile offspring of a male donkey and a female horse. Bioconservatives have worried about hybrids between humans and nonhuman animals. For example, an embryo resulting from a human egg fertilized by a nonhuman sperm would be a hybrid, as would an embryo created by inserting the nucleus of a fully human cell into an enucleated egg from a nonhuman species (since egg cells contain maternal DNA outside the nucleus). The latter has already been the subject of research in the United Kingdom, where, due to a severe shortage of human eggs for research the government body that regulates human embryonic research has permitted studies to create embryos by inserting human nuclei into cow eggs that have had their nuclei removed. The products of these procedures would not be permitted to develop beyond a few days in the laboratory, and they certainly wouldn't be able to produce some sort of science fiction version of a humanized cow. Rather, they are useful for providing important insights into human fertility and genetics and will be important in training embryologists.

More ethically and biopolitically charged chimera could be just over the horizon. Some new biological technologies present the possibility of recombining the genes that makes us specifically human with those of other primates. Say the question is about which genes give us our higher cognitive function or sense of self-awareness. The answers could lead to treatments for a multitude of cognitive and social disorders that are side effects of evolutionary changes in the human genome. But even well-targeted gene experiments might not be able to anticipate changes in other parts of an animal's genome, especially in the case of close evolutionary relatives like humans and chimpanzees.

Chimera are fairly recent, generally high-tech entrants to the stage of experimental biology. But since hybrids can sometimes be created through old-fashioned mating, they have long been just real enough to exercise the imagination, notably in H. G. Wells's *Island of Dr. Moreau*. Wells, it will be recalled, was an antivivisectionist who seemed to prefer animals to other people. In the 1920s, a Soviet scientist tried to breed a "humanzee," to fierce international objections. Among the objectors was Adolf Hitler, himself an antivivisectionist, who protested in *Mein Kampf* that "[t]he state is called upon to produce creatures made in the likeness of the Lord and not create monsters that are a mixture of man and ape." Chimera and hybrid fantasies are of course also a film industry staple. I knew that the term *genetic engineer* was penetrating the culture when it was used in the 1982 film *Blade Runner*, as part of a side plot in which a rather pathetic little fellow (the genetic engineer) amuses himself by creating sweet, compliant quasihumans as a hobby. The far less successful but visually compelling film *Splice* (2009) continues the tradition of depictions of arrogant scientists engaging in experiments with catastrophic consequences.

The next important political figure to refer specifically to hybrids in a major communication seems to have been George W. Bush. In his 2006 State of the Union address, Bush asked congress "to pass legislation to prohibit the most egregious abuses of medical research," including "creating human-animal hybrids. . . ." The president might have been inspired by then-Senator Sam Brownback of Kansas, who crusaded for years against human-nonhuman organisms. The senator's latest attempt, the Human-Animal Hybrid Prohibition Act of 2009, would prohibit the creation of a whole range of human-nonhuman organisms, most—but not all—of which are hybrids. The bill includes findings that human-nonhuman hybrids are "grossly unethical" because they "blur the line between human and animal, male and female, parent and child, and one individual and another individual." Brownback also emphasizes that human-nonhuman hybrids are a threat to human dignity and to "the integrity of the human species."

Brownback underscored that his new bill sought to prohibit only hybrids, thus allowing medically useful chimera-creations, such as pig heart valves. But the terms of his draft bill would have prohibited certain kinds of non-hybrid chimera as well. Most broadly, the bill would have banned "a human embryo into which a non-human cell or cells (or the component parts thereof) have been introduced to render the embryo's membership in the species *Homo sapiens* uncertain." Mixing human and nonhuman cells in a single *embryo* yields a chimera, not a hybrid. Hybrids comprise only those organisms that incorporate two or more distinct genomes in a *single cell*. This slippage is telling, and it threatens to hobble critical research. How many human cells can a chimera or hybrid organism have before its membership in the human species is "uncertain"? And what relevance ought we to attach to the developmental stage (embryo, fetus, born organism) of such organisms?

Although the Brownback bill never got out of committee, at the state level there has been some success on this track. In 2009, Louisiana governor Bobby Jindal signed into law a bill that was similar to Brownback's, prohibiting human-nonhuman hybrids. The Louisiana bill would forbid the intentional creation of a human embryo by any means other than fertilization of a human egg by a human sperm as well as "a nonhuman life form engineered so that it contains a human brain or a brain derived wholly or predominantly from human neural tissues." If the bill goes into effect, violation of the law would be considered a felony. Other states with strong bioconservative movements are attempting to impose bans on these kinds of organisms. For instance, a bill passed by the Ohio senate lists eight kinds of "human-animal hybrids" that would be outlawed, including normal human embryos into which animal cells are inserted, zygotes formed of one human and one animal gamete (for instance a human egg fertilized by animal sperm), animal eggs with an implanted human nucleus, and any "nonhuman life form engineered such that it contains a human brain or a brain derived wholly from human neural tissues." One supposes that, having been elected governor of Kansas in 2010, Mr. Brownback will put forward a similar bill in that state.

Rhetoric about human-animal hybrids simultaneously exploits both populist suspicion of hubristic scientists and reservations about capitalism. In 2010 one cultural conservative candidate for a U.S. Senate seat had claimed that "American scientific companies are crossbreeding humans and animals and coming up with mice with fully functioning human brains." Actually, what is being propagated in such statements is not human mice, but confusion. There is no denying that the hybrid issue excites broad public concerns about human hubris. The economist and activist Jeremy Rifkin has said of the creation of chimera, “One doesn’t have to be religious or into animal rights to think this doesn’t make sense. It’s the scientists who want to do this. They’ve now gone over the edge into the pathological domain.”

Crossing Boundaries

According to some etymologists, the words *hybrid* and *hubris* derive from the same Greek word, denoting a wanton act, an outrage, a mongrel—in effect, a crime against nature. Whether or not the etymology is accurate, the association of the two words goes to the heart of the concerns of those who, like Brownback, worry that any science or technology that undermines species boundaries calls into question what it means to be human. Species identification here is elevated to moral necessity. Again, biopolitics often reduces to worries over our place in the natural order, the great chain of being.

Cultural bioconservatives are especially worried that chimera and hybrids can blur species boundaries. The idea of a species is actually a fairly complex one, subject to various definitions, and scientific research has called into question whether the concept of species can, in fact, be fixed. Some would say that species are distinct if they don’t interbreed, but such a “definition” is suspiciously circular. Humans, of course, are also defined by species membership. In biopolitical discourse, what counts about these concepts is not their precise scientific justification but the fact that they represent a certain moral dimension that has to do with status in the natural order. But, as we’ve seen before, the great chain of being isn’t what it used to be. Deciding what the links are in the new chain of being and where we human beings fit will be a battleground of biopolitics.

Apart from President Bush’s remarks, the Brownback bill, and activity in a few states, chimera and hybrids have largely flown under the national political radar. President Bush’s bioethics council didn’t object to chimera-based medical research but drew the line at mixing human-nonhuman eggs and sperm. Any born products of such experiments, they argued, would put society in the difficult position of deciding whether a humanzee, for example, is more human or chimp. But a

United Kingdom House of Commons committee pointed out that such creatures would actually be less human than a human embryo, thus making research on them less objectionable than research involving embryos. They rejected worries about blurring species lines. But the fact that the United Kingdom has permitted the creation of human-animal hybrid embryos for research seems to have fueled the antihybrid forces in the United States.

This focus on species membership and particularly human uniqueness can be seen in familiar battles over abortion, where opponents have long argued that membership in the human species is sufficient to trigger the full panoply of legal rights. It is easy to see why blurring seemingly fixed species boundaries would therefore be a threat to the values these anti-abortion advocates hold dear. Debates over chimera and hybrids build on old divides but extend them into new contexts and against new ends. In this iteration, opponents of chimera-making cast themselves as protectors of the "human genome"—although the "human genome" is a statistical construct at best. Another notable aspect of both Brownback's bill and the state laws is their focus on the inviolability of the human brain. A similar bill up for consideration in the Arizona state legislature defines "human-animal hybrid" to include any combination of human cells and "non-human" life forms that involves an embryo and a "non-human life form engineered so that it contains a human brain." It is noteworthy that although these bills are justified partly by public health worries about dangerous novel viruses that could leap from the chimera to humans, brain-based chimera are the focus—not the "humanization" of other laboratory animals. In this legislation the human brain is given a special legal status above other vital organs. Perhaps the reason is that for many moderns, the brain is viewed as the seat of the soul, rather than other historic candidates for such status such as the liver or the heart.

Life from Nonlife?

Synthetic biology (or "synbio") is a relatively new laboratory field that, like the other issues explored here, presses on our sense of the natural. Indeed, synbio involves creating new life forms altogether. The basic tools of synbio are standard biological parts—sets of genes and chromosomes with known and specific functions created in modern biology labs—that can be assembled to program cells and control an organism's functions. "Biobrick" parts are already widely available from commercial Web sites, and the relevant skills for creating them are known to any reasonably competent biology graduate student. Made-to-order life forms may someday be used to repair diseased tissues, identify biological processes that are precursors to disease, create new energy sources, destroy polluting environmental toxins, and countless other uses that could significantly improve the quality of human life. Synthetic biology is only in its infancy, and it is likely to be combined with nanotechnology to create entities that blend the mechanical and biological.

Taking advantage of rapid advances in gene sequencing, many are now learning the techniques as part of the "do it yourself" (DIY) biology movement in which lone biohackers as young as high school students do bioengineering projects. As described in one article,

> These biohackers build their own laboratory equipment, write their own code (computer and genetic) and design their own biological systems. They engineer tissue, purify proteins, extract nucleic acids and alter the genome itself. Whereas typical laboratory experiments can run from tens-of-thousands to millions of dollars, many DIYers, knowledge of these fields is so complete that the best among them design and conduct their own experiments at stunningly low costs. With adequate knowledge and ingenuity, DIY biologists can build equipment and run experiments on a hobbyist's budget. As the movement evolves, cooperatives are also springing up where hobbyists are pooling resources and creating "hacker spaces" and clubs to further reduce costs, share knowledge and boost morale.

The DIYers' are following a path laid out by some top bioengineers. In 2010, a major step forward for synthetic biology was reported by Craig Venter and colleagues, who assembled a long strand of DNA from basic biological parts and successfully "rebooted" a cell into which the synthesized DNA had been integrated. Does an experiment like Venter's turn human beings into gods, making life from dust, as described in Genesis? Whether or not synthetic biology brushes too closely to this theological boundary, scientists long ago ventured into that frontier. If one takes the formulation of Genesis as the definition of God-playing, that distinction appears to have first been achieved in 1828, just ten years after Mary Shelley's *Frankenstein* established the paradigm for worries about hubristic science. Friedrich Wohler, while teaching chemistry in Berlin, applied ammonium chloride to silver isocyanate to produce urea, the main nitrogen-carrying compound found in the urine of mammals. In doing so, he synthesized an organic substance from non-organic matter. *Life from dust.* It was something the chemistry of the day took to be impossible, assuming that life could only come from life. Wohler's modest experiment proved them wrong, utterly changed organic chemistry, and laid down a philosophical marker—or perhaps kicked one over.

In addition to its research promise, synbio nonetheless raises the specter of peril. Concerns about synthetic biology can be subdivided into two primary categories: extrinsic and intrinsic. Extrinsic objections tend to be more utilitarian in nature. Here, they primarily concern environmental hazards of an accidental release of man-made organisms that may turn out to be harmful and difficult to eradicate, or even the deliberate synthesis of treatment-resistant bacteria or viruses for use in biological warfare. So far the loudest outcries have come from those on the left who fear the loss of biodiversity if synbio agricultural products are introduced. Although there is a complex system of law, regulation, monitoring, and professional self-policing that addresses these worries, such worries cannot—and should not—be dismissed. International cooperation and continued scrutiny at many levels of government will be required, as technology will almost certainly rush ahead of current conventions. Researchers and their supporters should also seek innovative approaches for verifying the character and safety of new life forms created through synbio. The rise and fall of public interest in and support for gene therapy should serve as a cautionary tale: a single serious incident, such as the death of Jesse Gelsinger at the University of Pennsylvania in 1999 during a gene therapy trial, can have an enormous impact on public perception.

Many individuals across the social and political spectrum may also find synbio intrinsically objectionable, perhaps as a violation of "naturalness," a standard that is ambiguous enough to be valued by both progressives and conservatives. Biopolitical realignment around naturalness is a phenomenon that has become familiar in Europe in the context of genetically modified food, for example. Nonetheless, more research on synthetic biology has so far been found acceptable across a wide range. When President Obama's bioethics commission issued a report in 2010 recommending no new regulatory arrangements for synbio, there was hardly any objection on intrinsic moral grounds. However, the technology is still in its early days, and specific applications or experiments might well change the political landscape for synthetic biology.

Connecting the Dots

Neuroscience and disorders of consciousness, predictive disease testing, chimera and synthetic biology may seem relatively disconnected from one another. This impression is mistaken. At least where we consider the biopolitical debates that are likely to arise around these issues, we can discern common threads linking concerns about these three issues: preserving human dignity and "naturalness." To critics of the Schiavo decision, cases like hers involve a primary obligation to feed all vulnerable persons, trumping modern notions of personal autonomy. To withhold food and fluids from those who cannot feed themselves seems unnatural. In the future, though, a variety of decidedly "nonnatural" technological interventions will be insisted upon by those who object to the Quinlan framework. The knowledge that one is liable to develop a depersonalizing disease is in one sense unnatural; it strikes a tone of a godlike omniscience that previous generations lacked. The new Louisiana state law banning the creation of human-nonhuman hybrids appears in the section of the criminal code that addresses crimes against nature, including copulating with animals. Both chimera and synbio also raise the specter of deviation from the status quo of species boundaries. Synthetic biology, by seeming to create new life forms from whole cloth, may be among the most grievous threats to this sense of the natural.

For example, when *New York Times* columnist Maureen Dowd tried to clarify what is really at stake in the confused chatter about chimera research, she inadvertently showed exactly why bioconservatives are worried about the power of the new biology and also why they and bioprogressives and science libertarians are talking past each other. Dowd asked about "the nightmare scenario" in which a mouse with a human brain could start acting human. Dowd wrote, quite accurately, that Stanford's Dr. Irving Weissman, who asked for ethical advice about the idea of making such a mouse, "is sensitive to ethical questions and has tried to ensure that 'the nightmare scenario' won't happen." But what is worrisome to many bioconservatives, and discomfiting to some bioprogressives, is not that any particular scientist might try it, but that biologists have acquired this power in the first place. Ultimately, they fear, there is no guarantee that someone won't try it, taking advantage of the fact that progressing science itself blurs biological and moral boundaries.

The question of what is "natural" arises in another context as well. Reminiscent of Joseph Fletcher's work in the 1970s, contemporary transhumanist philosophers like John Harris argue that our ability to enhance ourselves is part of our very human nature, and we should embrace the potential for science and technology to aid radical enhancements of our bodies and minds. On the other side, as is often true in the new biopolitics, there is at least a superficial alliance of bioconservatives and some bioprogressives, as both believe that excessive focus on cosmetic surgery as a statement of power over oneself is ultimately harmful to women. But bioconservatives embrace an idea of a core human nature that is being transgressed by such "enhancements." As Leon Kass writes, "If we can no longer look to our previously unalterable human nature for a standard or norm of what is good or better, how will anyone know what constitutes an improvement?" The question is a good one and the logic seems unassailable, but there's vigorous disagreement about what that norm of human nature is. In any case, the idea of an "unalterable human nature" is not necessarily one that is embraced by bioprogressives, who have more pragmatic worries about the messages being sent to young women by an enhancement-obsessed culture.

The Shape of Things to Come

The boundaries of life and death and those between species are obvious sites of biopolitical dispute because they play into other cultural divides, especially the one concerning human reproduction. These boundary issues provoke fears about loss of respect for what it is to be human. But at least the options being entertained are all within the realm of biology. By contrast, the blurring of boundaries between human beings and machines has excited far less attention but in fact might be far more perilous. One difference is that the earliest examples seem so modest and, at least in medical terms, so beneficial: cochlear implants, cardiac pacemakers, biomechanical joints and limbs, and, perhaps someday, devices like corneal implants and the long sought-after but elusive totally implantable artificial heart. There is no doubt that clinical medicine will soon offer a panoply of devices developed by biomedical engineers, possibly including nanotechnologies.

Few find these innovations objectionable in themselves, though there are important ethical issues at the research phase, especially about risks to the first patients. More striking are the implications of devices that will tie brains and nervous systems to machines. Again, the early phases could present great medical benefits, as in the case of people with quadriplegia. Experiments are being conducted that tie patients' neural impulses directly to computers so that they can manipulate a cursor, operate light switches, drive a wheelchair, and even maneuver an artificial arm and hand. In another arena, neuroscientists are working on brain implants that show promise in treating mental illnesses, mainly depression, that have not responded to standard therapy. Military planners have shown interest in noninvasive magnetic devices that seem to enhance learning and memory. Placed in helmets along with sensors to detect mental states like alertness, they might someday be able to stimulate a warfighter when a dangerous level of fatigue is indicated.

At this point some eyebrows might be raised at the possibilities inherent in such technologies, like the prospect of a new arms race or the imposition of new burdens on soldiers. However, there is probably not enough concern to stimulate major political reaction. More provocative questions are looming just around the corner. Some neuroscientists and computer scientists have speculated about the prospects for going well beyond treating psychiatric disorders or lifestyle enhancements like improving learning or attention. They speculate about external memory storage: literally downloading information directly into a port attached to the brain. They even speak of uploading unneeded information or memories to systems where they could be accessed at will. If this ever becomes technically feasible, it could provide an answer to the question of whether there is a separate self or, as David Hume thought, our sense of self is but a series of ideas with only the illusion of selfhood.

In other words, in our biopolitical preoccupation with biological breakthroughs we might be allowing history to effect a sort of sleight of hand, where the real action (and dangers) are in neuroscience rather than, say, genetics or stem cells. For nearly twenty years, a few computer scientists, philosophers, and technologists have entertained the idea of the "singularity," roughly defined as the point at which computers build other computers to answer questions that we aren't even able to formulate. Considering the astonishing growth of computing power, such a moment seems likely to be reached over the next decade or so, and many believe that in some ways we have already passed that point. An example is the control of financial or air traffic systems that are well beyond the capacities of human brains to simulate. Some transhumanists, who are not on the whole a pessimistic lot, believe that the singularity presents unprecedented risks that threaten human survival, even over the next century.

If exotic neurostimulation proves safe and feasible, alarms about the dangers of traversing natural human limitations will be sounded among bioconservatives; meanwhile, "green" progressives will warn about unfair access to neuroenhancements that could aggravate prevailing social inequalities. But will brain modification ever become the rallying cry that biotechnologies have become, exciting popular concern? If not, and if our existential survival is truly threatened by superintelligence beyond our control, the new biopolitics will have failed to organize human beings in ways that enable them to adapt to the world that neuroscience may now be creating. The result could look very much like that depicted in the last scenes of Steven Spielberg's 2001 film *A.I.*, in which that collection of abstract qualities we know as humanity survives as uploaded into silicon-based machines rather than as embodied in biological creatures.

The motivating idea of biopolitics has been the fear of biology without humanity. The converse, humanity without biology, might rather be what we should worry about.

CHAPTER SEVEN

IN DEFENSE OF PROGRESS

LIKE A CHARACTER IN A RUSSIAN NOVEL, AMERICA HAS FROM THE beginning been swept up in, and partly defined itself through, a fervid romantic triangle. Its two suitors: science and progress. Eager to court both, we Americans are especially liable to overlook our beloveds' shortcomings and exaggerate their virtues. We look to science for the growth of knowledge, new sources of wealth, and betters ways of living. Science is a fertile partner but full of surprises. Often our expectations of science are too great; the results are reliably plentiful but they can be disconcerting. If science is a blind date, progress is one that doesn't always show up. And when it does, it can be hard to recognize.

General Electric long touted one of the most effective slogans in the history of advertising: "Progress is our most important product." The slogan worked not only because of its appeal to America's sunny side, but also because GE's corporate spokesman, Ronald Reagan, exuded confidence in a perpetual morning in America; surely there was never a better minister for America's wedding to progress. Brought to you directly from the *Mad Men* era, the slogan also presupposed an unabashed and thoroughly presumptuously positive notion of progress. Here is an excerpt from Reagan's paper for GE's national sales meeting in 1961:

> At General Electric, you know, "PROGRESS IS OUR MOST IMPORTANT PRODUCT". But, all progress must have a starting point. All great human enterprises begin at a frontier. The pioneering of hostile

> lands . . . the mushrooming of the Atomic Age with its mixed blessing of destruction and construction . . . the rolling back of space frontiers beyond the planets to the stars . . . and the beginning of the information-handling revolution. ALL these had—a common starting point—a frontier of PROGRESS." (Emphases in the original.)

The alignment of message and messenger was just about ideal.

Thus it may come as something of a shock to be reminded that, in the words of the philosopher Nick Bostrom, "When headed the wrong way, the last thing needed is progress." Bostrom's observation would surely be embraced by all who give the matter some thought, especially those neoconservatives for whom Reagan became an iconic figure. Bostrom's simple dictum reminds us that progressives of all stripes need to avoid valorizing progress in the abstract. America's progress narrative is associated with a frontier, which is in turn linked to the wilderness and the West. Such is the power of the frontier image in the American mind, from Frederick Jackson Turner to Teddy Roosevelt to Vannevar Bush to JFK to Ronald Reagan, who appropriately enough earned his political spurs first as a Hollywood cowboy. Historically, Americans assumed that any venture that involved pushing into uncharted territory was progressive.

This much can be agreed upon regardless of one's place on the political map: Americans long for progress. One reason that Americans today have fallen into one of the country's periodic funks of self-doubt is that no compelling frontier has been offered that could justify change as progress. For progressives, however, the notion of a frontier as a new stage for the continuing American adventure is only the beginning; since the late nineteenthth century, they have insisted on material improvements in the conditions of life that could lift all boats, not just the yachts of the privileged few. Here, the idea of progress in our national mythos meets that of the American dream. Though wealth might be a bar to entry into the kingdom of heaven, in America the pursuit of happiness through earthly riches is perfectly acceptable, even desirable, so long as there is a sense of equality of opportunity. We need new frontiers that present new opportunities for wealth as well for the fun of conquering them. They provide a sense of national purpose and the promise—or at least the reasonable prospect—that our children may have a better life.

For progress of this sort to be a meaningful rallying cry for constructive social action, it must be associated with certain criteria. Following Bostrom's warning, how do we avoid progress in the wrong direction? The wilderness is uncharted territory; we need some lodestars. For the sake of argument, suppose we continue to accept the idea of progress according to Enlightenment values: improvement of the conditions of life through technology based on new knowledge and the reduction of superstition, intolerance, and cruelty. The point is not simply to clear the forest as we begin to stake a claim in the frontier, but to build a homestead, literally to make a homeland of this wilderness. Because the nature of progress is emergent, there is always going to be a measure of intuition in sensing whether, at first glance, the direction in which we are being led is the manifestly superior one. But gradually, evidence can be gathered and practices modified in light of that evidence; this is the public policy process.

Without governance, or what Foucault called "governmentality," there can be no such process. Policy making for social improvement does not require a particular conception of the sovereign or the state, or even a particular governance arrangement. It does, however, depend on a more or less coherent public voice. John Dewey understood a public to be created when citizens come to realize that they are being harmed or wronged by conditions beyond their control. Various types of negative conditions stimulate the creation of a public. Some of these problems proceed mainly from the natural environment, some from human creation, and some in combination, like climate change. One reason for the failure of totalitarian societies is that, in autocrats' self-protective need to create a false public imposed from above, they also stifle an authentic public voice from which the state could have learned about the conditions that fatally undermine its dictatorial authority. But not only closed societies can thwart attention to the public and its problems. Dewey worried that even in an open society many factors impair the public's ability to assume a coherent self-awareness and public voice. The distractions of corporate power, special interests, the media and communications technologies, and selfishness have been vastly amplified since World War II. Dewey would have considered our divisive, instantaneous, and relentless cable news, talk radio, and blogging culture as both the sources and evidence of a dispirited and inchoate public, one that is diverted from contemplating solutions to serious problems. But he would also have seen these media as presenting opportunities for new forms of public deliberation about matters of political importance.

Science and Progressives

The encounter between Benjamin Franklin and Edgar Allan Poe presented in the dueling epigraphs of this book represents competing and remarkably persistent strains in America's body politic: strains revisited in the new biopolitics. And what better, more original, or more contrasting representatives of their views could any nation boast? While much of Franklin's life was a joyous celebration of the material and intellectual prospects for the new nation, Poe's no less creative life was far shorter, troubled, and erratic. Franklin found love, recognition, and appreciation wherever he went; Poe hardly anywhere. And where Franklin envied future generations of Americans for the vistas of new scientific knowledge he could but dimly imagine, Poe mourned the impending, science-driven loss of the heart's credulity, of delightful reveries of gods, nymphs, and elves, an undoing of the spirit expressed in the purplest of prose. And when Poe did labor to write a scientific manuscript called *Eureka*, he was, to put it charitably, misinformed about the basics.

Was Poe right? Has science torn from us "the summer dream beneath the tamarind tree"? Though such charges are commonly made, the evidence of the last two centuries is not persuasive. Both popular and high culture continue to flourish; there is no indication that those societies that are oriented to science and technology are less culturally productive and vigorous. It is rather traditional and hidebound societies that are unable to revel in the unbound, in many cases because they are prevented from doing so by authorities that are explicitly or implicitly theocratic and conservative. The human need for reverie and romance appears to be far deeper than any science can wash away. In fact, considering only works like *Avatar* (and without making any aesthetic judgments), experimental science has, if anything, stimulated remarkable elaborations of ancient archetypes.

Franklin's remarks appear in a letter to Joseph Priestley in 1780, and Poe's lament was written in 1829; not, it seems, a promising trajectory for "progress" in the sense of optimism about science. The difference is more than a matter of personality, partly explicable by external events and partly by the intellectual movements of the times. Franklin was writing in the flush of revolutionary excitement as a new world was dawning for the former colonies; Poe's America was suffering a letdown from the founders' visionary leadership and engaged in the unglamorous work of nation-building. There was also the small matter of just holding the Union together. Factionalism and sectionalism had taken hold of American politics, with the aging John Adams lamenting that the new leaders lacked the vision of his generation. A familiar feature of generational change is the sense that conditions have deteriorated since the founding era. The Hebrews identified the Fall of Man, and the Greeks mourned a Golden Age. St. Augustine sought to resurrect the greatness of the disciples. The second and third Puritan generations felt that giants were standing on their shoulders. Poe and Mary Shelley both identified themselves with the Romantic movement, a reaction to European wars and political upheaval that celebrated a seemingly lost sense of natural purity and freedom. Both *Frankenstein* and the "Sonnet To Science" reflect the Romantics' fascination with folklore and fairy tales. The modern notion of progress as a journey toward successive improvements of the human condition is an exception to the historical rule, which was generally characterized by a sense of loss, decay, or even decadence.

This longing for a former, purer time in which good and evil are less ambiguous and full expression can be given to beauty has elements in common with the American neoconservative movement, which appeared after disillusionment with politics and in the wake of horribly destructive conflicts. These new conservative writers are motivated by a strong sense of morality and, especially under the influence of philosophers like Heidegger, the sense that something fundamental and mysterious has been lost in humanity's self-understanding. In both cases, science and technology are viewed as partly the causes and partly the consequences of these losses.

Nonetheless, bioprogressives should recognize the comprehensiveness of this cultural conservative biopolitical narrative. Whatever else may be said about it, it boasts a certain internal logic and consistency, is relatively straightforward in its defense of human dignity, and reaches deeply into the values of a society that is oriented to science and technology. The neoconservative bioethics narrative appeals to the sincerely held convictions of many about the ways that technology can undermine moral values cherished by conservatives and progressives alike. For these cultural conservatives, the Enlightenment has succeeded only too well. As Leon Kass wrote in his major work, *Toward a More Natural Science*, "The chickens are coming home to roost. Liberal democracy, founded on a doctrine of human freedom and dignity, has as its most respected body of thought a body of thought that has no room for freedom and dignity." Yet Kass goes on to assert that his position "does not arise from hostility to science."

This oppositional view between science and the Enlightenment on one hand and freedom and dignity on the other is a misreading of the history. Just as problematic, it leaves us in despair of the future. The core conviction of progressivism is that the institutions of governance (including but not limited to government itself) must adapt to technological change and that in doing so, the conditions of human life can be improved. There is no inconsistency between progressivism and sustaining a role for traditional values in this adaptive process. Progress is not a radical break from the past, but rather on a continuum with it. Improvements build on previous experience and currently available assets. In this sense, progressive social action is an effort to find and, if needed, institutionalize continuity. But modern left liberals vary in their views about the risks and benefits of the new biology and therefore disagree about what counts as progressive social action in the realm of biopolitics. Among them are bioprogressives who often end up sounding like libertarians and greens who often end up sounding like neoconservatives, but each for very different reasons from the individualists on the one hand and the social conservatives on the other. One reason for this pair of anomalous results is that, as Marcy Darnovsky notes, the biopolitical left has inherited the autonomy orientation of modern bioethics. In terms of political philosophy, the emphasis on patient autonomy fits well with "procedural" liberalism, which emphasizes the fairness of the process itself, a concept of rationality, and neutrality about what counts as the good life. In their study of modern liberals and conservatives, several psychologists have found that liberals tend to emphasize fairness, while conservatives rely on a wider set of moral foundations. The Anglo-American legal system tends to default to procedural liberalism, as does much of the rest of public life in modern

America, theoretically leaving individuals to identify and pursue their vision of the good life.

According to green progressives and cultural conservatives, procedural liberalism in politics leaves the public unsatisfied. As Darnovsky puts it, "many Americans care deeply about the moral and spiritual aspects of political issues and want public policies and electoral campaigns to address them." It also dangerously removes progressives from substantive arguments about a moral vision for the country. In Michael Sandel's words, it leaves that territory to "narrow, intolerant moralisms. Fundamentalists rush in where liberals fear to tread." Sandel advocates a substantive liberalism that emphasizes a sense of civic obligation and encourages the development of a virtuous moral character. Darnovsky and Sandel agree that, in spite of their good intentions about tolerance, bioprogressives err if they cede a vision of values like human dignity to cultural conservatives.

The reintegration of human dignity into a cohesive bioprogressive vision might seem like a stretch in light of the recent heated exchanges during the embryonic stem cell research debate and other such biopolitical controversies, but it is a worthy and important project. And it might not turn out to be such a philosophical heavy lift for the science-sympathetic left. After all, dignity is a key element of virtually every important human rights convention, with documents on the subject drafted by sophisticated modern liberals who believe that the international rule of law requires cooperation among sovereign states. Are technology and human dignity compatible? Can a new biopolitics of progress be crafted? We are only at the very beginning of the effort to sort out these questions. But as John Dewey liked to say, experience has a way of solving philosophical problems. Inspiration can often be found in a particular person's experience. My candidate for such inspiration is a young woman named Cody Unser.

Racing Ahead

Americans are by no means the only people who find fast cars fascinating, but there's a good case to be made that the Indianapolis 500, like the World Series, the Super Bowl, and the Triple Crown, is among those events that symbolize the national spirit. Indy cars are generally the most technologically sophisticated of racing cars. These are American-style "open wheel" vehicles with axles that extend the tires outside the body of the car. Setting them apart from stock cars or even sports car racing, which uses equipment related to what might be seen on the street, open-wheel cars have a single seat and are built for speed. Unlike all other classic American sporting events, one family is uniquely identified with open-wheel racing and the Indianapolis 500. This family is the Unsers: Al Unser, his brother Bobby Unser, and Al Unser, Jr. Between them, they have won the race nine times.

Race cars are now firmly lodged as among the very symbols of American moral virtue, technical virtuosity, and driving ambition (pun intended), but it was not always so. At the end of the nineteenth century, many harbored grave doubts about the implications of new forms of transportation, fearing that they would loosen social bonds and challenge traditional morality. This was, after all, the Victorian era. (Considering the significance of drive-in movies and backseats for a generation of young Americans, perhaps they were not wholly wrong). It fell to a devout Christian and innovative beekeeper, A. I. Root, of Medina, Ohio, to signify that neither automobiles nor "aeroplanes" were incompatible with sound American values. Root made this point in one vivid demonstration that he reported in his widely read journal, *Gleanings in Bee Culture*. Root described his decision to buy a 1903 Oldsmobile Runabout and drive it on two hundred miles of miserable roads to Dayton, where on September 20, 1904, he watched Wilbur Wright fly his heavier-than-air flying machine in a complete circle for the first time. The man who became wealthy with his innovative design for man-made beehives is an unlikely all-but-forgotten hero in the continuing American struggle with technology and values.

It was while thinking about Americans and fast cars, A. I., Root, and America's apprehensions about technological innovation that I came to know Cody Unser. As the daughter, granddaughter, and niece of legendary Indy 500 champions, she is a kind of American royalty. The philosopher Hegel thought of Napoléon as a "world historical" individual whose life both summed up all that had come before and provided the germ for the next stage of history. We moderns no longer appreciate that kind of metaphysical grandiosity, but Hegel's notion does capture the way that some lives are lived at a number of historical intersections. Cody's life is one of those.

For Cody Unser, the idea of the body politic has special meaning. When Cody was twelve years old, she was stricken with transverse myelitis. Paralyzed from the chest down, she faced the usual challenges of adolescence as well as learning how to live in a wheelchair and manage a variety of medical problems related to her disability. She committed herself to staying fit and as healthy as she could and to finding a way to walk again. In conversation, Cody likes to disabuse those of us who are temporarily able-bodied of the idea that people with disabilities, especially but not only young people, are uninterested or incapable of intimacy. Her straightforwardness is perhaps characteristic of a new, no-nonsense generation.

While barely in her twenties, Cody became an advocate for embryonic stem cell research. She was inspired by her own experience, the stem cell debate that raged about her while she grew into adulthood, and the person she calls her "superman," Christopher Reeve. Of course, there are many patient advocates; what sets Cody apart is the combination of her age and the scholarly approach she has taken to the issues. As an undergraduate at the University of Redlands in California, Cody designed her own major in "biopolitics." For her senior project, she taught a course on biopolitics to thirteen other Redlands students. Before I learned about Cody, I had never heard of a self-designed biopolitics major. In 2010, Cody moved to Washington, D.C., to pursue a graduate degree in public health.

Cody is exquisitely aware that her heritage as an Unser puts her in a unique position to communicate with those who might have deep moral reservations about the life sciences. I first learned about her when I watched her testify before a U.S. Senate subcommittee, just as I was thinking about this last chapter. This could be a pretty unnerving experience even for someone much older, and I found her remarkably poised under the circumstances; it was clear that for someone her age, she'd had a lot of experience in a public role. Yet when we talked a few weeks after the senate hearing, she told me she's still working on how best to use her voice. "As an advocate, I'm trying to understand more about how to use my personal story and bring it together in a more powerful way," she says. "I don't want to just use my story; I want to back it with facts and the truth. And what these [embryonic stem cells] are and what the science is."

I wondered how a person of tender age but profound experience who had reflected on and lived through the new biopolitics—and who had a big stake in one of its core issues—would respond to the ideas in this book. In her situation, Cody has an intense relationship with technology. She's not only dependent on a wheelchair but also uses special exercise equipment like a mechanical standing frame. Cody has a passion for scuba diving, another technologically mediated way of relating to the world; in the water, she can experience her body through movement made possible by its natural buoyancy.

So Cody relates to a far wider range of technologies for more needs and purposes than most of us. What lessons has that taught her? "People who walk don't necessarily think of their shoes as technology as much as I do my wheelchair. There's a difference there that is interesting," she said. "I rely on [technology] just to get around, from point A to point B. So I've learned so much about how technology influences science and health. . . . there are two different avenues it seems . . . to think of the body not only as biology but also as engineering."

And as the physical sciences converge with the life sciences, what do we gain, and what are we at risk of losing? I found that Cody's concise response hit the mark:

> I think science gets at the core of our human vulnerability. Both in a negative and a positive [way]. Science has evolved so fast and it freaks people out because we don't want to lose what makes us human. Science brings about cures in the medical world. Stem cell research one day will be able to treat disease and disability. But I don't think that we will ever lose what makes us human. Science will never be able to push us back that far.

Fixed on Science?

Earlier in this book, I described the elite debate about artificial reproductive technologies in general, and human embryonic stem cell research in particular, that have become part of American politics. The protagonists in that elite debate have this in common: they see the underlying dynamic as a confrontation between science sympathizers and science skeptics. For some, the outcome of this debate will turn on the most important questions of human values and political philosophy.

For example, the conservative thinker and former Bush administration adviser Yuval Levin has objected to some progressives' charge that the American conservative movement has engaged in a "war on science." This claim, he says, tells us little about modern conservatives but "does tell us something about the American left and its self-understanding." Levin concedes that the left has indeed become "the party of science" but contends that progressives' fixation on science has been a mixed bag, leading to "the gruesome experiment in applied social science called communism."

The charge that there is a progressive "fixation on science" is one that has frequently been sounded by neoconservatives, especially in reference to the reaction of scientists and their allies to the stem cell controversy. In a 2002 speech, President George W. Bush said, "The powers of science are morally neutral—as easily used for bad purposes as good ones. In the excitement of discovery, we must never forget that mankind is defined not by intelligence alone, but by conscience. Even the most noble ends do not justify every means." Often, however, cultural conservatives argue that science is anything but morally neutral; specifically, science, especially in its most technological form, contains an implicit value system that tends to favor domination over nature, which I have argued is a remarkably Marxist view. At other times, cultural conservative worries seem to be less about science itself and more about the arrogance of the scientists who practice it, as reflected in President Bush's statement.

Calling on his experience as a White House domestic policy staffer, Levin argues that an orientation to a certain notion of progress as defined by scientific advance often combines with a "crisis" mentality with regard to social problems. The powerful desire to find a fix for human ills like disease, he says, is especially vulnerable to the sense of unfairness through which we perceive tragedies that befall the innocent, and that could well entangle each of us too. Levin views this fight against disease as understandable and admirable but also somewhat misplaced, as though suffering can always be avoided if we only find a technical solution. Like many cultural conservatives, Levin sees the impetus behind advocacy of embryonic stem cell research as an acute case of overreaching scientists. "Most opponents of embryo research, after all, have not suggested that we abandon the quest for cures and relief. They seek rather to pursue it in a moral way." Such a formulation leaves the reader to infer that those on the other side of the argument are not as concerned with the morality of the means to scientific breakthroughs. The left, he says, is especially prone to a certain blindness with regard to the power of science, staking far too much faith on its inherent beneficence.

Of course, science cannot provide an account of human dignity or equality, nor does it purport to do so. As we have seen, the measure of science's contributions to human dignity and equality lies not in its means but in its method, for the method of science is the great leveler. It implicitly promotes a meritocracy of testable ideas rather than a notion of a priori insight attributable to those who happen to occupy positions of authority. The results of inquiry and demonstration may be inconvenient, but that is beside the point. Levin argues, for example, that in strictly biological terms, it is precisely human *inequality* that is becoming more evident through the study of genetics, and this is a problem for the left.

It is hard to know what to make of this provocative observation. Levin is right that modern genetics is turning up individual differences, but his use of the word *inequality* is misleading. Equality is a political, not a biological concept. Though we do differ in our genes, my genes and yours operate in a complex interplay with the environment described by a field called phenomics. Another growing field, epigenetics, studies the way environmental influences may be passed on to offspring by modified genes that nonetheless maintain their basic DNA sequence. The uniqueness of phenomic and epigenetic processes for any single individual may be such that earlier worries about total genetic control are beyond the reach of any biotechnology; there just might be too many variables to track.

In any case, without the Enlightenment values that have fostered experimental science, no one would even be worried about human equality as a moral principle. The notion that humanity is universal in spite of differences in appearance, culture, language, or religious beliefs is still too often only honored in the breach; in no way was it a commonly held assumption before the eighteenth-century Enlightenment philosophers. Since then, political rights, including the right to worship or not as one prefers, have matured into the concept of human rights. Meanwhile, we have become inured to the fact that scientific ways of thinking have yielded a rich conception of equality by valorizing systematic demonstration of statements about the world rather than taking mere authority for granted.

Science Surprises

There is no doubt that the new biology has disquieting implications that will continue to stir the biopolitical pot. But I think these will make themselves felt in more surprising ways than can now be predicted, and they won't necessarily center on familiar issues like the moral status of the human embryo, the politics of heredity, or even the commodification of human beings and their parts. Consider the story of the Havasu 'Baaja tribe or, as they are more commonly called, the Havasupai. With only about 650 members in 2010, the tribe won a momentous lawsuit for control over their genetic material. In 1882, the U.S. Government declared the tribe's Grand Canyon land to be a national park, and the Havasupai were confined to a small area at the bottom of the canyon. After one hundred years of wrangling, the government restored 185,000 acres to the tribe but by then traditional hunting, fishing, and farming had long been replaced by tourism as the tribe's main means of earning a living. Along with the tourists and visitors came all manner of new nonnative food and drinks. The Havasupai found themselves ravaged by type 2 diabetes. In 1990 three scientists at Arizona State University started a research project to better understand the diabetes epidemic. This project was done with the support of the U.S. National Institutes of Health and with the approval of the university's ethics review board and of the seven-member council of the Havasupai tribe. The investigators took at least 400 blood samples from 180 tribal members to determine whether genetic factors might be putting some members of the tribe at risk of type 2 diabetes.

But the genetic studies revealed little about a proclivity to diabetes. By 1992 the researchers expanded their studies to include genetic and medical records analysis of inbreeding, schizophrenia, migration history, and genealogy—without the explicit consent of those who had supplied the original blood samples. More than a dozen papers were published on these topics. It was only when a tribal member attended a talk at Arizona State University about some of the non-diabetes work that the tribe became aware of the additional studies. They were angry and filed suit against the researchers, the institutional review board, and the University Regents. In April 2011, the Arizona State University's Board of Regents agreed to pay $700,000 to forty-one of the tribe's members, return all blood samples, provide scholarships, and help build a new health clinic for the impoverished Havasupai. Although no liability was acknowledged, the settlement suggests that, in the eyes of the law, research participants can be wronged when they are not fully and precisely informed about the way their DNA might be used.

One reported concern of the Havasupai was that the genetic studies also showed that their ancestors had migrated from Asia, almost certainly via the Bering Strait, conflicting with their tribal creation story that has them formed by gods from the dust and waters of their canyon. In the Havasupai creation legend, the first human being is a woman (logically enough), who then bears the progenitors of the rest of humanity. Ever since this primordial event, their responsibility has been to serve as protectors of the Grand Canyon. Passing this unifying legend down from one generation to the next has been part of the function of tribal elders, so to challenge it is to undermine family bonds. What the new biology gives with one hand (answers and possible treatments for genetically linked diseases) it takes away with another (a sense of tribal identity and purpose).

Genetics is also revealing that more contemporary Americans have Native American ancestors than is generally appreciated. The presence of Native Americans always caused trouble for the Puritan worldview, resulting in some odd theological contortions to explain their presence in the new Eden. Little comment on the execrable treatment of these peoples should be required; their forced removal to the West anticipated twentieth-century ethnic cleansing. By putting facts like genetic origins before us, the new biology is creating new biopolitical power relations, even for those most excluded from the American narrative. The example of the encounter between Havasupai Indians and the sophisticated laboratory science of a modern university is a reminder of the key to the new biopolitics: the struggle for control over the new biology and the information and potential power it represents. The Havasupai story is one that should humble bioprogressives like me. It turns worries about the meaning and control of the new biology on its head. Unlike eugenics, the issue is not that of passing identifiable traits from one generation to the next, but rather concerns the cultural effects of minimizing the richness of an "unscientific" worldview. It is not about destiny, but about origins. The Havasupai case demonstrates the volatility, depth, and unpredictability of the new biopolitics.

Fire from the Left

Conservatives are not the only ones who criticize the left for an uncritical pro-science tilt. Since the 1970s the most influential attack on science has come from within the academic left itself, in the form of postmodernism. Drawing on philosophers and historians like Karl Popper and Thomas Kuhn, the postmodernists noted that because scientists formulate statements that are derived from previous experience, there is an unavoidably subjective element in science. This subjective element (whatever that is taken to mean) has been the target of a vast academic industry that appeals to analytic categories and language drawn from Marxism, psychoanalysis, anthropology, sociology, and economics, all intended to reduce science to a social construction.

That science requires generalizing from previous experience is not a matter of dispute. Often these generalizations turn out to be false. So "all swans are white" disappeared from the medieval logic texts when a black swan was discovered. What is interesting, and what sets science apart, is precisely that its statements can be disconfirmed and systematically revised in accordance with observation and experiment. Of course, all experience is in some trivial sense "subjective." To assert that science is a social construction is to say nothing interesting about the validity of the statements formulated as a result of experience by a community of inquirers. Nuclear physics, genomics, and learning theory have all been developed by talented individuals working in teams and across the ages; that they all had psyches and points of view doesn't make bombs, DNA, or literacy any less substantial entities, nor does it make the entities' manipulation for blast waves, energy production, antibiotic resistant organisms, vaccines, indoctrination, or education any less possible.

But suppose for the sake of argument that, postmodernists notwithstanding, the left is the party of science—then where stands the right? The goal of modern conservatives must not be to become the anti-science party, a self-defeating position if ever there was one. American conservatives of the older sort, before neoconservatives had fully captured the energy of the right, seemed to have a different attitude toward science. Writing in 1980, the conservative sociologist Robert Nisbet asserted that "No single idea has been more important than the idea of progress in Western civilization," naming the worth of economic and technological growth as one of the crucial premises of the idea of progress. It is not at all clear that technology would have such a high standing in a set of premises formulated by the neoconservative commentators today, or at least not without much qualification; it is also unclear if they would so wholeheartedly endorse Nisbet's defense of the idea as one of continuing value. Recent neoconservatism has rather defined itself as the party of ethics, seeking to rein in the scarier implications of the new biology. Besides the dubious and simplistic "ethics versus science" frame (is science not at least in part a moral practice, and doesn't ethics require argument and evidence?), there is in the contemporary bioconservatism of the right an implicit appeal to popular notions of a scientific Faustian bargain. For the sake of an illusory progress, we necessarily sell our souls. As it is undeniable that no one can guarantee the outcome of an experiment, fears about the consequences of change can never be wholly resolved. But a liberal democratic society has nothing to fear and everything to gain by fostering a scientific attitude. Timothy Ferris has well described liberal democratic government as "an endlessly changing mosaic of experiments, most of which partially or entirely fail. This makes the process frustrating but generates its strength."

The term *experiment* has come to be most widely applied to America itself. If ever there was a social science experiment it has been the question, in the words of one noted commentator, whether a nation so conceived and so dedicated can long endure. In modern terms, America is not a done deal. Yet the lessons of experimentation can only be incorporated into experience if there is a spirit of open inquiry. The fundamental value that drives science is not material or technological progress, but the value of openness to evidence, including especially evidence that is incompatible with our prejudices. This point was well appreciated by the American founders. For them, as it must be for us, America is a question—a hypothesis about self-government—not an answer. Although Marxist-Leninists touted a "scientific socialism," when the evidence of the catastrophic consequences of their system became clear, it was ignored or rationalized or suppressed for the sake of power. The harsh lesson is that openness to inquiry in itself is not enough. Strong civic institutions are required for a society to absorb and implement the lessons of experience. When Benjamin Franklin was asked what kind of government the constitution had created, he famously replied, "A republic, if you can keep it."

Whither Biopolitics?

While it is impossible to predict with confidence the ultimate significance for the future of American politics of the new biology and the biopolitics it has spawned, a few conclusions may be drawn with some level of confidence. One is that it is in the interest of cultural conservatives that biopolitics does not assume a prominent role in American politics. In spite of their ambivalence, Americans have consistently though often fitfully embraced the sometimes disconcerting implications of science and new technologies once convinced that they were within what William Carlos Williams called "the American grain." The notion that experimentation is largely laudable is, as I have noted, deep in the American character. A willingness to experiment, to risk encountering evidence that confirms one's own fallibility, reflects what William James called a "will to believe" about the future. Even the revolutionary generation had deep misgivings about the plausibility of betting their lives and fortunes on a republic that enshrined personal religious freedom in a "bill of rights," but they thought the experiment worth the gamble. Although experimental biology has led to an acute cultural crisis about science, at the end of the day, the convergence of the life sciences with the physical sciences and engineering may blunt the force of that rift. Meanwhile, the strongest card neoconservatives can play in biopolitical controversies is that of isolating them from other policy questions by emphasizing biological advances as uniquely threatening to human values.

As John Dewey observed, a splintered public is an obstacle to gathering political support for measures that advance the material conditions of life and the common good. Nonetheless, these measures require science. Scientific innovation is the natural ally of progressive philosophy because it is the key to material progress itself. And if the idea of America as an experiment is more than a rhetorical conceit or a convenient narrative fable, the idea of scientific inquiry must be an important part of our civic inheritance. Contrary to much modern cultural conservative thought, Elysian fields are not the logical conclusion of the progressive celebrants of science; a gradually better life is all that is sought. Nor should progressives cede the moral high ground. As *The Grapes of Wrath* attests, human dignity is as populist an idea as there is. As the new bioethical issues emerge and flow into the new biopolitics, progressives need to link technological advance to individual opportunity and a greater sense of social solidarity. Bioprogressives should acknowledge that not all applications of new knowledge are acceptable and should urge a shared sense of responsibility for the direction of science. At the same time, a robust appreciation for the importance of innovation must be central to any progressive philosophy.

There are those who seek more than gradual improvement, but perfection itself. For many transhumanists, human perfection is a reasonable goal. Though this may seem outlandish, transhumanists may take comfort in another very American tradition called "perfectionism." The Puritans sought perfection in their relation to God, though they proved intolerant of others' attempts to find their own version. After the Articles of Confederation proved too weak to hold the country together, the constitution adopted in 1787 specified "a more perfect union." Christian perfection of the spirit was the explicit goal of the mid-nineteenth-century American Holiness movement, followed a few decades later by "muscular Christianity" in the YMCA and the Fellowship of Christian Athletes, which introduced physical education as essential to moral perfection.

Ronald Reagan made John Winthrop's stirring vision of a "city on a hill" a key trope in his campaigns for the presidency. The phrase, which appears in Jesus's Sermon on the Mount, was one that provided the Puritans with spiritual sustenance in their arduous adventure. But as historians of colonial America have observed, the early settlers were more impressed with what they found than with what they brought. That sense of wonder at the new world and the vast and mysterious land that lay before them, and a requisite continuing sense of moral purpose, were captured in a sermon delivered by the Rev. Samuel Danforth in 1670 entitled "A Brief Recognition of New-Englands Errand into the Wilderness." Although Augustine might have found his city on a hill in the clouds, for Americans without an errand into the wilderness, there can be no city on a hill. Bioprogressives would do well to embrace and encourage that sense of adventure and discovery; they are, in any case, its natural heirs.

One Body Politic

For all its ambiguity, the idea of progress that is so important to Americans must be placed in the context of the Enlightenment, and progressivism as a political philosophy only has meaning if it appears in the context of some national story. America's civic narrative is the most vivid political realization of the Enlightenment linkage between progress and science. A dynamic, living progressivism requires a historical challenge. For the progressives of the late nineteenth and early twentieth centuries, the challenge was the effects of industrialization; for today's progressives, the challenge is globalization: how to preserve the promise of the American dream for all, of a level playing field and equality of opportunity, in a very different world from that in which America's civic narrative of frontiers for discovery, adventure, and prosperity flourished. The globalization of science creates new opportunities for knowledge as well as new struggles for control over the results of that knowledge. Nowhere is that struggle for control more evident than in the new biology.

Stirring as it is, the American narrative that gives flesh to the idea of progress is stained with the blood of those who were left out. John Winthrop's errand has too often been used as a formula for a gloating "American exceptionalism." Like persons, history judges nations by their acts, not their words. America has indeed been blessed, but as the ancient Hebrews learned, the blessings of providence must finally be earned. African slaves were led into the wilderness in chains, if they were fortunate enough to survive the Middle Passage. For them, the city on a hill was hardly even a promise in the clouds. As little consolation as it might be, at least some moral progress can be claimed in the belated recognition that the New World did not mean freedom from oppression for all who arrived.

Happily, however, one example of the power of the new biology is the details it can provide for an argument about human solidarity. Every living human being's mitochondrial DNA comes from one African woman, and all male Y chromosomes come from one man. Sarah Tishkoff, my colleague at the University of Pennsylvania, spends much of her time in Africa collecting blood samples. She and her team have constructed remarkable genetic maps that show ancient human origins, inheritance patterns, linguistic developments, and critical migration paths starting in southern Africa and exiting the continent at the Red Sea tens of thousands of years ago. Tishkoff's research has all sorts of implications for medicine, including the development of powerful and safer drugs that are suited to one's genetics. Yet the same basic rules of biology explain and govern our differences. Her work shows that, in the end as well as in the beginning, we are all members of the same body politic.

REFERENCES

Introduction

15 *harms outweigh the benefits*. National Science Board, *Science and Technology Indicators 2010*. http://www.nsf.gov/statistics/seind10/.

15 "*to society's well being*." The Pew Research Center for the People & the Press, *Public Praises Science; Scientists Fault Public, Media* 2009. http://people-press.org/report/528/.

16 *facts they track*. The spokesman was Rick Weiss, former *Washington Post* reporter.

16 "*religious right's point of view*." "Evolution, Big Bang Polls Excluded from NSF Report," *Science Insider*, April 8, 2010. http://news.sciencemag.org/scienceinsider/2010/04/evolution-big-bang-polls-omitted.html.

16 *matters of controversy*. And controversies about new technology often miss the mark. In 2010 conservative Virginia state legislators worried that implantable microchips, which use G.P.S. technology to track Alzheimer's patients who tend to wander, might be the "mark of the beast" warned of in *Revelations*. In their fundamentalist zeal the Virginia elected officials who worried about Biblical prophecy missed a truly concerning social question raised by implanted microchips: the prospect that someday everyone's physical location could instantly and constantly be recorded and accessible.

17 *opposed human embryo cloning for research*. Genetics and Public Policy Center, "Cloning: A Policy Analysis," 2005 http://www.dnapolicy.org/images/reportpdfs/Cloning_A_Policy_Analysis_Revised.pdf.

17 *small majority opposed it*. Yuval Levin, "Public Opinion and the Embryo Debates," *The New Atlantis*, Spring 2008. http://www.thenewatlantis.com/publications/public-opinion-and-the-embryo-debates. Arguably the wording of the relevant question was rather loaded: "It is unethical to destroy human embryos for the purposes of research because doing so destroys human embryos that are human beings and could otherwise have developed and grown like every other human being."

17 "*human embryos used in the research*." Pew Forum on Religion and Public Life, 2008. "Declining Majority of Americans Favor Embryonic Stem Cell Research." http://pewforum.org/docs/?DocID=317.

18 "*total rejection of scientific methodology*." John Evans, University of California at San Diego, unpublished manuscript, 2010.

18 *French philosopher and historian Michel Foucault*. A number of Foucault's

works illustrate his view of biopolitics and biopower, including: *The Order of Things* (New York: Vintage Books, 1970); *Discipline and Punish: The Birth of the Prison*, Trans. Alan Sheridan (New York: Vintage Books, 1977); and *The History of Sexuality, Volume 1: An Introduction*. Trans. Robert Hurley (New York: Vintage Books, 1978).

18 *bodies that we call populations*. Although the elements of Foucault's thinking described here have been a source of inspiration for part of the argument in this book, I don't endorse all of Foucault's view about science, or what are often taken as his views, including that science has been mainly part of a conspiracy of hegemonic exploitation. It seems to me that the role of science in political history is more subtle than was captured by the fashionable academic slogans of the 1970s and 80s.

19 "*distinct from the right of the sovereign*." Jason Robert, M.A. Thesis: "Biotechnologies of the Self: The Human Genome Project and Modern Subjectivity," McMaster University, 1996. Robert is one of a handful of academic bioethicists who have taken Foucault's work seriously, or who have attended to biopolitics more generally. See also Jeffrey P. Bishop and Francois Jotterand, "Bioethics and Biopolitics," *Journal of Medicine and Philosophy*, 31:205-212, 2006.

19 "*birthrate, longevity, race*." Cited in Mitchell Dean, *Governmentality: Power and Rule in Modern Society* (Thousand Oaks, CA: SAGE, 1999), p. 99.

19 "*basic components of life*." Majia Holmer Nadesan, *Governmentality, Biopower, and Everyday Life* (New York: Routledge, 2008), p. 2.

19 "*biological century*." I believe this term was first used by Gregory Benford in an article in *Engineering & Science* in Spring, 1992. At the risk of being a lot more pedantic than I want to be in this book, I believe the grammatically correct expression is "the century of biology." Still, the point is clear enough.

20 "*biopolitical practices and discourses*." Paul Rabinow, "Artificiality and Enlightenment: From Sociobiology to Biosociality," in Mario Biagioli (ed.), *The Science Studies Reader* (New York: Routledge, 1999).

21 *indirectly to electrical theory*. Edison seemed to believe that, in his case, too much conscious focus on theory would impair his remarkable intuitive powers. Referring to the formula for voltage, Edison recalled, "At the time I experimented I did not understand Ohm's Law. Moreover, I do not want to understand Ohm's law. It would prevent me from experimenting." Cited in Harold Evans, *They Made America* (New York: Little, Brown), 2004, p. 192. Interestingly, Morse also didn't seem to know much of the relevant theory of the day with regard to batteries for electricity; if he had he might have saved himself much time.

21 "*invention without innovation is a pastime*." *Ibid*. pp. 6-7.

23 "*scientists, politicians and market libertarians*." Jeremy Rifkin, "This Is the Age of Biology," *The Guardian*, July 28, 2001. http://www.foet.org/global/BC/This%20is%20the%20age%20of%20biology.pdf.

24 *compensating organ donors*. Frances Kissling and Sally Satel, "How Marion Barry Could Help the Organ Shortage," *Washington Examiner*, March 30, 2009. http://www.washingtonexaminer.com/opinion /columns /OpEd-Contributor /How-Marion-Barry-could-help-the-organ-shortage-42116237.html #ixzz11hMRb3ub.

26 "*guanine substitution*." Rabinow, *Ibid*.

Chapter 1

30 *represented by theEnlightenment.* For a summary of the evidence see Timothy Ferris, *The Science of Liberty* (New York: Harper Collins, 2010). However, the historian Niall Ferguson argues that the rapid economic growth of the previous couple of hundred years that is partly due to technologic change was responsible for the horrific violence of the 20th century. See Niall Ferguson, *The War of the World* (New York: Penguin, 2006).

30 "*new, infant republic.*" Richard Nisbet, "Idea of Progress," *Online Library of Liberty*, http://oll.libertyfund.org/?option=com_content&task =view&id=165 &Itemid=259#lf-essay004lev2sec02.

33 "*inheritance from our forefathers.*" Edmund Burke, *Reflections on the Revolution in France* (New York: Pearson Longman, 2006), p. 40.

33 *must be falsifiable.* The philosopher was Charles S. Peirce, of whom more below.

36 "*over the same country.*" Alexis de Tocqueville, *Democracy in America* (New York: A. S. Barnes & Co., 1851), p. 337.

38 "*portion of human knowledge.*" Alexis de Tocqueville, *Democracy in America.* William James was happy to embrace this stereotype in his version of pragmatism, which stressed that the meaning of an idea lies in its practical consequences.

41 "*systematic investigation.*" Charles Rosenberg, *No Other Gods: On Science and American Social Thought* (Baltimore: John Hopkins, 1976), p. 18.

42 "*believed about nature.*" Christopher P. Toumey, *God's Own Scientists: Creationists in a Secular World* (New Brunswick, NJ: Rutgers University Press, 1994).

42 "*not one dollar voted.*" Cited in Rosenberg.

43 *all that surrounds it.* Daniel Dennett, "Universal Acid." *Darwin's Dangerous Idea: Evolution and the Meanings of Life* (New York: Simon and Schuster, 1995).

44 *attempting to know the world.* Charles S. Peirce, "How to Make Our Ideas Clear" and "The Fixation of Belief," originally published in *Popular Science Monthly*, November 1877 and January 1878, respectively.

45 *manner of science.* It's important to appreciate that Peirce was no atheist. The insights that make it possible for us to hypothesize and theorize more or less accurately have part of a natural order and a higher mystery that we can never quite comprehend. ". . . [T]he discoveries of science, their enabling us to predict what will be the course of nature, is proof conclusive that though we cannot think any thought of God's, we can catch a fragment of his Thought, as it were." In Charles Hartshorne and Paul Weiss, eds., *Collected Papers of Charles Sanders Peirce*: Volumes V and VI (Cambridge, Massachusetts: Harvard University, 1965), p. 372 of Vol. 6.

46 *public policy process.* Though James himself detested TR's trademark, swaggering militarism.

48 *his cousin, Franklin Delano.* However, FDR was not a trustbuster, a basic tenet of classical American progressivism.

51 *is often compared.* Francis S. Collins, Michael Morgan, Aristides Patrinos, "The Human Genome Project: Lessons from Large-Scale Biology," *Science* 300(5617): 286 – 290. 3003.

51 *ionizing radiation*. These projects also raised questions about the ethics of human experiments, which I describe in *Undue Risk: Secret State Experiments on Humans* (New York: Routledge, 2000).

52 *moon landing*. Pew Research Center for the People and the Press, "One Small Step No Longer Seen as Such a Giant Leap for America," July 15, 2009.

53 *societal goals and values*. Daniel Sarewitz, *Frontiers of Illusion* (Temple University Press, 1996).

53 *to fund science*. As a foreign science attaché at an embassy in Washington, D.C., once said to me, "It's funny but I've never met a scientist who had enough money."

54 *for the US economy*. Research!America, "Research: An Economic Driver," 2010, http://www.researchamerica.org/uploads/factsheet14economic.pdf.

Chapter 2

55 *influence over behavior*. People like me who grew up in small towns and found their way to cities as adults often delight in the anonymity found in bustling urban centers. With that anonymity comes a sense of individuality and the opportunity to reinvent oneself. But with the Web and social networking sites like Facebook one may no longer be able to escape one's past. The whole world becomes a small town in which an established identity through gossip is inescapable and the sense of individuality might be undermined.

56 "*or of oneself*." Cited in Nikolas Rose, Pat O'Malley, Mariana Valverde, "Governmentality," *Annual Review of Law and Social Science*, 2:83-104 (Volume publication date December 2006).

57 "*governing too much*." Nikolas Rose, Pat O'Malley, Mariana Valverde, "Governmentality," *Ibid*. p. 84.

58 *authoritative history*. The word is used to refer to policies and practices in different cultures and countries. See Mark Adams, ed. , *The Wellborn Science: Eugenics in Germany, France, Brazil, and Russia* (New York: Oxford University Press, 1990).

58 *horrors of Nazi Germany*. Paul Lombardo, "Looking Back at Eugenics," in *A Century of Eugenics in America: From the Indiana Experiment to the Human Genome Era*, ed. Paul A. Lombardi (Bloomington: Indiana University Press, 2010).

58 *occasion for biopolitics*. Though he had little to say about eugenics per se, Foucault proposed a course at the Collège de France on the knowledge of heredity in the nineteenth century, "starting from breeding techniques, on through attempts to improve species, experiments with intensive cultivation, efforts to combat animal and plant epidemics, and culminating in the establishment of a genetics whose birth date can be placed at the beginning of the twentieth century."Cited in Ladelle McWhorter, "Governmentality, Biopower, and the Debate over Genetic Enhancement," *Journal of Medicine and Philosophy*, 34:409–437, 2009.

61 *concentration camp experiments*. For a fascinating account of Brandt's life and career as one of Hitler's favorites, see Ulf Schmidt, *Karl Brandt – The Nazi Doctor: Medicine and Power in the Third Reich* (London: Continuum, 2007).

62 *the most useful of all citizens*. Cited in Daniel Roche, *France in the Enlightenment* (Cambridge, MA: Harvard University Press, 1998), p. 420.

64 *new practices*. Claude Bernard, *An Introduction to the Study of Experimental Medicine*, 1865. First English translation by Henry Copley Greene (Macmillan & Co., Ltd., 1927; reprinted in 1949).

64 *rearranging the world*. George Bernard Shaw, *The Quintessence of Ibsenism: Now Completed to the Death of Ibsen* (New York: Brentano, 1913).

64 *and professional training*. Daniel Kevles, *In the Name of Eugenics: Genetics and the Uses of Identity* (Alfred A. Knopf, 1985).

65 *their "natural ability."* Statistics is also central to modern genetics, which looks for statistical correlations, this time between genes and health, called Genome Wide Association Studies or GWAS.

65 *reduction of "undesirables."* These inferences of course overlooked the advantages of money, class, and what we would today call social networks.

65 *should be avoided*. It should be said on Galton's behalf that he was a first-rate scientist who made many contributions to statistics, among other fields, and was opposed to Lamarck's theory of inherited characteristics. But, like Spencer, he was an enthusiast for marriage between persons of high social rank.

66 *upheaval and violence*. Garland E. Allen, "Is a New Eugenics Afoot?" *Science* 294:59–61, 2001 .

66 *tides of the ocean*. Cited in Rosenberg, 1976.

67 *more fortunate fellows*. William Castle, *Genetics and Eugenics* (Cambridge, MA: Harvard University Press, 1916). I am grateful to Paul Lombardo for bringing this remark to my attention.

67 *homeless, tramps and paupers*. Mark H. Haller, *Eugenics: Hereditarian Attitudes in American Thought* (Rahway, NJ: Rutgers University Press, 1963).

67 *legalized eugenics*. Paul Lombardo, *Three Generations, No Imbeciles* (Baltimore, MD: Johns Hopkins, 2008).

67 *imbeciles are enough*. Buck v. Bell, 274 U.S. 200, 47 S.Ct. 584, 71 L.Ed. 1000 (1927).

68 *University of Heidelberg in 1936*. Glen Yeadon and John Hawkins, *The Nazi Hydra In America: Suppressed History of a Century* (Joshua Tree, CA: Progressive Press, 2008).

68 *control of human evolution*. Garland E. Allen, "Is a New Eugenics Afoot?" *Science* 294:59–61, 2001.

69 *prophetic novels*. H. G.Wells, Anticipations of the Reaction of Mechanical and Scientific Progress Upon Human Life and Thought (New York and London: Harper & Brothers, 1902).

69 *the White House*. Jacqueline M. Moore, *Booker T. Washington, W.E.B. Du Bois, and the Struggle for Racial Uplift* (Wilmington, DE: Scholarly Resources, 2003).

69 *eugenic interpretation of genetics*. At the same time, many Nazi leaders were restrained in their enthusiasm for nuclear physics because its leading figure was a Jew, a prejudice that helped to save the world from the catastrophe of a Third Reich with an atomic weapon.

71 *mentally or physically wanting*. Yuval Levin, *Imagining the Future* (New York: Encounter Books, 2008), p. 101.

71 *competitive society*. Michael Sandel, *The Case Against Perfection: Ethics in the Age of Genetic Engineeering* (Cambridge, MA: Harvard University Press, 2007), pp. 77–78.

72 *could have been.* Allen Buchanan, Dan W. Brock, and Norman Daniels, "Why Not the Best?" *From Chance to Choice: Genetics and Justice* (New York: Cambridge University Press, 2000), p. 159.

73 *even in wealthy countries.* The average cost of IVF is currently $12,000. Having PGD can tack on an additional $3,000 or so.

74 *body mass index, for example.* Sha, B.Y., Yang, T.L., Zhao, L.J., Chen, X.D., Guo, Y., Chen, Y., Pan, F., Zhang, Z.X., Dong, S.S., Xu, X.H., Deng, H.W., "Genome-Wide Association Study Suggested Copy Number variation May Be Associated with Body Mass Index in the Chinese Population," *Journal of Human Genetics* 54(4):199–202, 2009.

74 *extra or missing.* Interview with Nancy Spinner, "Unraveling our own Code," *Science Progress*, April 3, 2008.

75 *metabolites in urine.* Eriksson, N., Macpherson, J.M., Tung, J.Y., Hon, L.S., Naughton, B., Saxonov, S., Avey, L., Wojcicki, A., Pe'er, I., Mountain, J. "Web-Based, Participant-Driven Studies Yield Novel Genetic Associations for Common Traits," *PLoS Genetics* 6(6), 2010.

76 *as well as others.* Quoted in Thomas Goetz, "Sergey Brin's Search for a Parkinson's Cure," *Wired,* July 2010. http://www.wired.com/magazine/2010/06/ff_sergeys_search/all/1.

76 *useless information. Der Spiegel* Interview with Craig Venter, "We Have Learned Nothing from the Genome," *Der Spiegel Online*, July 29, 2010 http://www.spiegel.de/international/world/0,1518,709174,00.html.

77 *biopolitical contention.* Ronald Bailey, "I'll Show You My Genome, Will You Show Me Yours?" *Reason,* January 2011. http://reason .com/archives/2010/12/13/ill-show-you-my-genome-will-yo.

78 *musical gene.* My Gene Profile. http://www.mygeneprofile.com/talent-test.html.

78 *race or ethnicity.* Sandra S.J. Lee, Joanna Mountain, Barbara Koenig, et al., "The Ethics of Characterizing Difference: Guiding Principles on Using Racial Categories in Human Genetics," *Genome Biology* 9(7), 2008. http://genomebiology.com/2008/9/7/404.

78 *generalizations about race. Ibid.*, "Our hope is that scientific data about human genetic variation might undermine spurious popular beliefs about the existence of biologically distinct human races and beliefs that support racist ideologies."

78 *of race and genetics.* For the foreseeable future, though, pharmaceutical firms and some patients may be among the most immediate beneficiaries of the new computational genetics. My University of Pennsylvania colleagues have found that some people have a specific gene that raises their risk for heart disease; these results also suggest that drugs could be designed to change those individuals' genetic destiny. The technology is Genome Wide Association Studies. The main funder of this important workis President Obama's economic stimulus package. If it does lead to new drugs, this kind of discovery will in turn rely on transfer to the private sector in order to manufacture enough medication to save countless lives. No simple walls can be built between new forms of inhumanity and new opportunities for human flourishing. Therein lies the perpetual paradox and challenge of the new biology.

Chapter 3

82 *point is to change it.* Karl Marx. "Theses On Feuerbach," In *Marx/Engels Selected Works*, Vol. 1 (Moscow: Progress Publishers, 1969), p. 15, 1845.

82 *more comprehensively, the life sciences.* The life sciences comprise all the biologically-based fields of knowledge. I use the terms biology and life sciences interchangeably unless otherwise indicated.

83 *and other disciplines.* Joel Cracraft, "A New AIBS for the Age of Biology," *Bioscience 2004.* http://www.aibs.org/bioscience-editorials/editorial_2004_11.html.

84 *those of our own.* Elizabeth A. Phelps and Lauren A. Thomas, "Race, Behavior, and the Brain: The Role of Neuroimaging in Understanding Complex Social Behaviors," *Political Psychology* 24(4):747-758, 2003. http://www.psych.nyu.edu/phelpslab/papers/role%20of%20neuroimaging%20in%20understanding%20complex%20human%20behaviors.pdf.

85 *cephalic organization.* Jay Schulkin, *Naturalism and Evolution: Pragmatist Legacies and Prophecies*, forthcoming 2011.

86 *often referred to as "cloning."* Even the name of this technology has been a matter of intense political debate. From the biologist's point of view cloning is an imperfect term for the maneuvers in question because it refers to so many laboratory techniques. They prefer the more descriptive "somatic cell nuclear transfer," but the use of that phrase has been challenged by opponents of the research as too neutral for most people to understand the stakes involved in this particular type of cloning.

86 *for the genetic material.* Watson, J.D. and Crick, F.H.C. "A Structure for Deoxyribose Nucleic Acid." *Nature* 171(3): 737–738, 1953.

87 *too much of a machine.* Paul Ramsey, *Fabricated Man: The Ethics of Genetic Control*, (New Haven, CT: Yale University Press, 1970).

88 *our inherent potential.* The oft-used term "open future" was coined here. Feinberg, J., "The Child's Right to an Open Future," in *Whose Child? Children's Rights, Parental Authority and State Power*, eds. William Aiken and Hugh La Follette (Totowa, NJ: Rowman and Littlefield, 1980), p. 124–153.

89 *Luddite rejection of cloning.* International Academy of Humanism, "Statement in Defense of Cloning and the Integrity of Scientific Research," May 16, 1997.

90 *depth of his shallowness.* Leon Kass, "Science, Religion, and the Human Future," *Commentary*, April 2007. An earlier version of this paper was published as "Triumph or Tragedy: the Moral Meaning of Genetic Technology," Appendix F in Francis Fukuyama and Caroline Wagner, *Information and Biological Revolution: Global and Biological Revolutions: Global Governance Challenges – Summary of a Study Group*, Rand 2000.

90 *any scientific inquiry.* Steven Pinker, Letter to the Editor, *Commentary*, July/August 2007.

94 *converse among themselves.* The problem is not limited to scientists in relation to non-scientists. We have reached a point at which highly specialized scientists who are formally in the same field or academic department, or even working on closely related problems do not necessarily understand their colleagues' work.

95 *prominently displayed.* G. Drori, J. Meyer, F. Ramirez, E. Shofer, *Science in the Modern World Polity* (Palo Alto, CA: Stanford University Press, 2003).

95 *first time in 2008*. Science-Metrix, "30 Years in Science: Secular Movements in Knowledge Creation," 2010 (http://www.science-metrix.com/30years-Paper.pdf).

95 "*multiple institutions*." Maltsev, N. "Computing and the 'Age of Biology,'" *CTWatch Quarterly*, 2(3), August 2006. http://www.ctwatch.org/quarterly/articles/2006/08/computing-and-the-age-of-biology/.

96 "*view the data*." *Ibid*.

97 *paranormal phenomena*. Heather Ridolfo, Amy Baxter, Jeffrey Lucas, "Social Influences on Paranormal Belief: Popular Versus Scientific Support," *Current Research in Social Psychology* 15(3), 2010.

98 *"cloud" computing systems*. I wrote this book using such a system for word processing. Instead of storing my chapter drafts only on my computer hard drive or on a disk or memory key, I saved each draft in a system that uploaded it to software distributed throughout the Web. I could open it from any computer in the world that had Web access.

Chapter 4

101 *by* The Washington Post. Rick Weiss, "Conservatives Draft a 'Bioethics Agenda' for the President," *The Washington Post*, March 6, 2005. http://www.washingtonpost.com/wp-dyn/articles/A15569-2005Mar7.html.

104 *like a normal embryo*. Wilmut, I., Schnieke, A.E., McWhir, J., Kind, A.J., & Campbell, K.H.S., "Viable Offspring Derived from Fetal and Adult Mammalian Cells." *Nature* 385: 810–813, 1997.

105 *As the first cloned*. Scientists prefer the phrase "somatic cell nuclear transfer" to "cloning" to describe the process pioneered at Roslin because cloning refers to so many copying techniques in biology. But critics of embryonic stem cell research charge that the use of cloning is less technical and is one that non-scientists have become familiar with. It's another example of how important language is in communicating science to the public.

105 *to become gods*. For an excellent summary of the views of Fletcher and Ramsey as they persisted in the more recent cloning debate see Courtney Campbell's commissioned paper for the National Bioethics Advisory Commission, "Cloning Human Beings: Religious Perspectives on Human Cloning."

105 *survives the procedure*. *Cloning Human Beings: Report and Recommendations of the National Bioethics Advisory Commission*, Rockville, Maryland: 1997.

106 *majority of nations*. United Nations General Assembly, "General Assembly Adopts United Nations Declaration on Human Cloning by Vote of 84-34-37." Press release. *Fifty-Ninth General Assembly*. March 8, 2005.

109 *the National Academies*. The National Academies are comprised of the National Academy of Sciences, the National Academy of Engineering, the Institute of Medicine and the National Research Council.

109 *for embryonic stem cells*. I co-chaired this committee from 2004 to 2005, when our report was published, and served on its successor advisory committee until it was dissolved in 2009.

109 *in* The New York Times. Nicholas Wade, "Group of Scientists Drafts Rules on Ethics for Stem Cell Research," *New York Times*, April 17, 2005. http://www.nytimes.com/2005/04/27/health/27stem.html.

113 "*develops by degrees*." Michael Sandel, "Embryo Ethics – The Moral Logic of

Stem Cell Research," *The New England Journal of Medicine* 351(3):207-209. July 15, 2004.

114 *intrinsic character*. Robert P. George and Patrick Lee, "Acorns and Embryos," *The New Atlantis*, Fall 2004/Winter 2005, pp. 90-100.

115 *separate 2007 papers*. Takahashi, K., Tanabe, K., Ohnuki, M., Narita, M., Ichisaka, T., Tomoda, K., Yamanaka. S. Induction of pluripotent stem cells from adult human fibroblasts by defined factors. *Cell* 131(5): 861-72, 2007. Yu, J., Vodyanik, M.A., Smuga-Otto, K., Antosiewicz-Bourget, J., Frane, J.L., Tian, S., Nie, J., Jonsdottir, G.A., Ruotti, V., Stewart, R., Slukvin, I.I., Thomson, J.A.. "Induced Pluripotent Stem Cell Lines Derived from Human Somatic Cells." *Science* 318(5858):1917–20, 2007.

118 *as many as 70*. "Graphic Feed: Cultivating Regenerative Medicine Innovation in China," *GloberHigherEd*, January 9, 2010. http://globalhighered.wordpress.com/2010/01/09/graphic-feed-cultivating-regenerative-medicine-innovation-in-china/.

118 *ambitious scientists*. A couple of days before Christmas 2005 I was asked to meet with several senior officials of the South Korean government at a very deserted National Academies building so that they could deliver their heartfelt apology for the incident. Apparently they were making the rounds in Washington to deliver the same message to various organizations. I was asked to meet them in my capacity as former chair of the Academies committee on stem cells. Their shame was obvious, deep, and quite moving.

118 *little internal debate*. Curiously that includes South Korea in spite of its influential evangelical Christian movement.

119 *spinal cord injuries*. Geron Corporation. Safety Study of GRNOPC1 in Spinal Cord Injury. Received October 6, 2010. http://clinicaltrials.gov/ct2/show/NCT01217008?term=GRNOPC1&rank=1.

120 *event of a spill*. Postscript to the *Chakrabarty* case: In theory the genetically modified bacteria that were patented could have been used to respond to the 2010 gulf oil spill. But the company where the bugs were developed, General Electric, didn't maintain the stored bacteria. For invention to matter discovery alone is not enough.

120 *traditional ideological spectrum*. The poignant story of Henrietta Lacks illustrates the social importance and widespread acceptance of privatizing the value of human tissues, but it is an occasion for the queasiness many of us feel about that practice. A thirty-one-year-old African-American mother of five, Lacks died in 1951, not long after she was diagnosed with cervical cancer. Many of the most important cell lines in medical science labs for the past sixty years were derived from the cancerous tissues of this long-deceased woman. These "HeLa" cells turned out to be remarkably prolific. For some still unknown reason they were able to be grow virtually indefinitely. These "immortal" cell lines have been used in research that led to the polio vaccine and were mass produced for research on cancer, radiation and toxin effects, AIDS, genetics and even to develop cosmetics. Science writer Rebecca Skloot pieced the story together in *The Immortal Life of Henrietta Lacks* (2010), noting that Lacks's family had only the vaguest information about what was done with her cancer cells or how they could be used without her or their permission; it was conventional at the time, though it does fuel race and class suspicion. Still more ironically, the Lacks family has been without the health insurance that would give

them access to many of the discoveries made with Henrietta's tumor cells. Today ethical standards require disclosure of the possible commercial value of tissues, and that the patient would not benefit from that value.

Chapter 5

122 *commodified, fragmented body.* Lesley A. Sharp, "The Commodifiction of the Body and Its Parts," *Annual Review of Anthropology* 29:287–328, 2000.

122 *necessary to therapeutic cloning.* Donna Dickenson, "Commodification of Human Tissue: Implications for Feminist and Development Ethics," *Developing World Bioethics* 2:55-63, 2002.

122 *alien to the laborer.* Karl Marx, op. cit., 1967.

123 *determines their consciousness .* Karl Marx, *Contribution to the Critique of Political Economy.* Moscow: Progress Publishers, 1977. http://www.marxists.org /archive/marx/works/1859/critique-pol-economy/preface.htm. Emphasis added.

124 *nature of modern man.* Irving Kristol, "Keeping Up with Ourselves," in *The End of Ideology Debate*, ed. Chaim Waxman (New York: Funk and Wagnalls, 1968). http://writing.upenn.edu/~afilreis/50s/kristol-endofi.html.

124 *as they wished.* Francis Fukayama, "A Milestone in the Conquest of Nature," in *The Future is Now*, eds. William Kristol and Eric Cohen, (New York: Rowman and Littlefield, 2002) 77–80.

124 *alienating it from them.* Karl Marx, "Excerpt Notes of 1844," in *Writings of the Young Marx on Philosophy and Society*, Eds. Lloyd Easton and Kurt Guddat, (New York: Doubleday and Company, 1967), 265–282.

125 *a 'posthuman' age.* Adam Wolfson, "Why Conservatives Care About Biotechnology," *The New Atlantis* 2:55–64, 2003.

126 *unpredictable and undependable.* Gertrude Himmelfarb, "Two Cheers (or Maybe Just One) for Progress," in *The Future is Now*, ed. William Kristol and Eric Cohen (New York: Rowman and Littlefield, 2002), 73–76.

126 *heaven on earth.* William Kristol and Eric Cohen, "Cloning, Stem Cells, and Beyond," in *The Future is Now*, edited by William Kristol and Eric Cohen (New York: Rowman and Littlefield, 2002). 297-305.

126 *What they want is – more.* Irving Kristol, *op. cit.*, 1968.

126 *virtue of human beings.* Eric Cohen, "New Genetics, Old Quandaries: Debating the Biotech Utopia," *The Weekly Standard*, 7(41), 2002. http://www .weeklystandard.com/Content/Public/Articles/000/000/001/132jmnjg.asp?pg=1

127 *my life for theirs.* "Live with Leon Kass," *The American Enterprise*, July/August 2006.

127 *of our will.* Adam Wolfson, *op. cit.*, 2003.

127 *alienation of man.* Eric Cohen, *op., cit.*, 2002.

127 *consequences on the other.* Paul Cella, "Technology and the Spirit of Ownership," *The New Atlantis* 9:55–64, 2005.

128 *bureaucratic middleman. Ibid.*

128 *paid wage labourers.* Karl Marx, "The Communist Manifesto," in *The Marx-Engels Reader*, ed. Robert Tucker (New York: W.W. Norton and Company, 1978(, 469–500.

128 *vast bureaucratic corporations.* Paul Cella, "Technology and the Spirit of Ownership," *The New Atlantis* 9:55–64, 2005.

128 *all our achievements. Ibid.*

129 *economics is a tool. Ibid.*

129 *not animals or machines. Beyond Therapy: Biotechnology and the Pursuit of Happiness.* The President's Council on Bioethics, Washington, D.C., October 2003.

129 *recommendation is never made.* Brandon Keim, "The Strange Saga of the President's Council on Bioethics," *GeneWatch* 17(3), 2004. http://www.gene-watch.org/genewatch/articles/17-3keim.html.

130 *mere money relation.* Karl Marx, "The Communist Manifesto," in *The Marx-Engels Reader*, Robert Tucker, ed. (New York: W.W. Norton and Company, 1978) 469–500.

130 *native biology.* Eric Cohen, "Biotechnology and the Spirit of Capitalism," *The New Atlantis* 12:9–23, 2006.

130 *will be unstoppable.* Leon Kass, "Preventing a Brave New World," in *The Future is Now*, William Kristol and Eric Cohen, eds. (New York: Rowman and Littlefield, 200), 219–241.

130 *'smart shoppers.'* Adam Wolfson, *op. cit.*, 2003.

130 *'commodities exchange.'* Eric Cohen, "Biotechnology and the Spirit of Capitalism," op. cit., 2006.

130 *is an illusion. Ibid.*

131 *in the first place.* Paul Cella, *op. cit.*, 2005.

131 *long known it.* Eric Cohen, "Biotechnology and the Spirit of Capitalism," *op. cit.*, 2006.

131 *body most especially. Ibid.*

135 "*dictates of science.*" Martin Heidegger, "The Question Concerning Technology," in *The Question Concerning Technology and Other Essays*, William Lovitt, ed. (New York: Harper Torchbooks, 1977) 3–35.

135 *to be manipulated.* Tabachnick, *op. cit.*, 2005.

135 *fertility treatment.* Leon Kass, "The New Biology: What Price of Relieving Man's Estate?" *Science* 174(4011):779–788, 1971.

135 *organ transplants.* Leon Kass, "Organs for Sale?" The Public Interest 107:65–86, 1992.

135 *human cloning.* Leon Kass, "The Problem of Technology," in *Technology in the Western Political Tradition*, Arthur Melzer, Jerry Weinberger, and M. Richard Zinman, eds. (Ithaca: Cornell University Press, 1993) 1–24.

135 *never encountered directly. Ibid.*

135 *application of method. Ibid.*

136 *full human life.* Eric Cohen, "The New Politics of Technology," *The New Atlantis* 1:3–8, 2003.

136 *reining it in. Ibid.*

136 *to live well.* "Live with Leon Kass," *op. cit.*, 2006.

137 *Judeo-Christian ethics.* Edward Ashbee, "'Remoralization: American Society and Politics in the 1990s," *The Political Quarterly* 71(2):192–201, 2000.

140 *technology and enhancement.* James J. Hughes, "Technoprogressive Biopolitics and Human Enhancement," in Progress in Bioethics: Science, Policy and Politics Jonathan D. Moreno and Sam Berger, eds. (Cambridge, MA: MIT Press, 2010).

140 *mental states and moods.* Nick Bostrom, "In Defense of Posthman Dignity," *Bioethics*, Vol. 19, 2007.

141 *potential bioterrorist.* G. Annas, L. Andrews and R. Isasi. "Protecting the Endangered Human: Toward an International Treaty Prohibiting Cloning and Inheritable Alterations." *American Journal of Law and Medicine* 28:2&3, 2002. p. 162.

141"*most dangerous idea.* Francis Fukuyama, "The World's Most Dangerous Idea: Transhumanism," *Foreign Policy* 144:42–43, September 4, 2004.

141 *needed to be eradicated.* Ronald Bailey, "Transhumanism: The Most Dangerous Idea?" Reason.com, August 25, 2004.

Chapter 6

143 "*tissues and cells.*" A good example is the history of anti-vaccination campaigns, which began in the 18th century when British churchmen argued that smallpox was a divine punishment and that therefore inoculation is a sin. (Why the divine was capable of designing smallpox but not a resistant strain went unexplained.) More recently, anti-vaccination campaigns have focused on allegations that vaccination against deadly diseases like diphtheria and whooping cough themselves pose unacceptable risks. These allegations have never stood up to the evidence, though their potential popular appeal is very worrisome to public health officials. In any case the motivations for such dangerous claims need to be understood as originating in longstanding suspicion of science, expertise and governance.

143 *ruling beliefs.* Leon Kass, *Toward A More Natural Science: Biology and Human Affairs* (New York: Free Press, 1985).

144 *reason of things.* Charles S. Peirce, *Collected Papers of Charles Sanders Peirce*, Charles Hartshorne and Paul Weiss, eds. (Cambridge: Harvard UniversityPress, 1965-68).

144 *free speech.* R. Alta Charo, "Politics, Progressivism, and Bioethics," in Jonathan D. Moreno and Sam Berger (eds.) *Progress in Bioethics: Science, Policy, and Politics* (Cambridge, MA: MIT Press, 2010).

145 . . . *midst of "embryoville."* Leon Kass, "Reflections on Public Bioethics: A View from the Trenches," *Kennedy Institute of Ethics Journal* 15:3 (September 2005).

150 *food and hydration.* Cruzan v. Director, Mo. Dept. of Health, 497 U.S. 261, 284 (1990).

151 *heretofore been envisioned.* Geert De Meyer, Fred Shapiro, Hugo Vanderstichele et al., "Diagnosis-Independent Alzheimer Disease Biomarker Signature in Cognitively Normal Elderly People," *Arch Neurol.* 67(8):949-956, August 2010.

153 *cross species boundaries.* Robert, J.S., Baylis, F. "Crossing Species Boundaries." *American Journal of Bioethics.* 3(3):1–13, 2003.

153 *Parkinson's Disease.* Bjugstad, K.B., Redmond, D.E. Jr, Teng YD, Elsworth, J.D., Roth, R.H., Blanchard, B.C., Snyder, E.Y., Sladek, J.R. Jr. "Neural stem cells implanted into MPTP-treated monkeys increase the size of endogenous tyrosine hydroxylase-positive cells found in the striatum: a return to control measures." *Cell Transplant.* 14(4):183–92. 2005.

153 *stroke.* Chu, K., Park, K.I., Lee, S.T., Jung, K.H., Ko, S.Y., Kang, L., Sinn, D.I., Lee, Y.S., Kim, S.U., Kim, M., Roh, J.K. "Combined Treatment of Vascular Endothelial Growth Factor and Human Neural Stem Cells in Experimental Focal Cerebral Ischemia." *Neuroscience Research.* 53(4):384–90, December 2005.

(Another investigation into the use of neural stem cells to improve brain function in stroke model.)

154 *studies of AIDS*. Namikawa, R., Kaneshima, H., Lieberman, M., Weissman, I.L., McCune, J.M. "Infection of the SCID-hu Mouse by HIV-1." *Science,* 23;242(4886):1684–6, December 1988.

154 *leukemia*. Kamel-Reid, S., Letarte, M., Sirard, C., Doedens, M., Grunberger, T., Fulop ,G., Freedman, M.H., Phillips, R.A., Dick, J.E. "A Model of Human Acute Lymphoblastic Leukemia in Immune-Deficient SCID Mice." *Science* 22; 246 (4937):1597–600, December 1989.

155 *'never go forward.'* Chimeras: Animal-Human Hybrids," PBS Online News-Hour, August 16, 2005. http://www.pbs.org/newshour/bb/science/july-dec05/chimera_8-16.html.

155 *human eggs for research. See* Human Fertilization & Embryology Auth., Press Release: "HFEA Statement on Licensing of Applications to Carry Out Research Using Human-Animal Cytoplasmic Hybrid Embryos," Jan. 17, 2008. http://www.hfea.gov.uk/418.html.

157 *predominantly from human neural tissues*. HB 2652 , State of Arizona House of Representatives, Forty-ninth Legislature, Second Regular Session, 2010. http://www.azleg.gov/legtext/49leg/2r/bills/hb2652p.htm.

157 *wholly from human neural tissues*. SB 243, State of Ohio Senate, 128th General Assembly, 2009–2010 http://www.legislature.state.oh.us/bills.cfm?I=128_SB_243.

157 *fully functioning human brains*. "Christine O'Donnell on Human Mice, Lying to Nazis, and the Women of Middle Earth," *Washington Post*, September 10, 2009. http://voices.washingtonpost.com/roughsketch/2010/09/christine_odonnell_on_human_mi.html.

158 *pathological domain*. "Animal-Human Hybrids Spark Controversy," *National Geographic News*, January 5, 2005. http://news.nationalgeographic.com/news/2005/01/0125_050125_chimeras.html.

159 *embryos for research*. "Hybrid Human-Animal Research Approved in the UK," *Science Daily*, January 18, 2008. (http://www.sciencedaily.com/releases/2008/01/080118102223.htm).

160 *boost morale*. Paraijata Mackey, "DIY Bio: A Growing Movement Takes on Aging," *H+ Magazine*, January 22, 2010 http://www.hplusmagazine.com /articles/bio/diy-bio-growing-movement-takes-aging.

161 *extrinsic and intrinsic*. I owe this analysis to Andrew Light.

161 *gene therapy trial*. Zallen, D.T. "U.S. Gene Therapy in Crisis," *Trends in Genetics*. 16(6):272–75, 2000.

161 *self-identified progressives*. Jesse Graham, Jonathan Haidt, and Brian A. Nosek, "Liberals and Conservatives Rely on Different Sets of Moral Foundations," *Journal of Social Psychology* 96(5): 1029–1046, 2009.

163 *won't happen*. Maureen Dowd, "Slouching Toward Washington," *The New York Times*, September 26, 2010.

163 *philosophers like John Harris*. John Harris, *Enhancing Evolution: The Ethical Case for Making Better People* (Princeton University Press, 2007). I owe this point to Rebecca Kukla.

163 *constitutes an improvement?* Leon Kass, *Life, Liberty and the Defense of Dignity* (San Francisco: Encounter Books, 2002), p. 132.

Chapter 7

168 *last thing needed is progress.* Nick Bostrom, http://www.nickbostrom.com/.

169 make a homeland of this wilderness. The recent appearance of the term "homeland" as a description of the geographic territory of the United States (the Department of Homeland Security, for example) is jarring for those of us who are accustomed to thinking of that as the term for places of ethnic origin, the "old country". Surveys of immigrants to the United States indicate that many individuals had no intention of remaining forever. One way to interpret the American project is the very effort to make this place our homeland, to create an American nation. But once this goal has been achieved the project itself is over, then what?

172 *hostility to science.* Leon Kass, *Toward a More Natural Science* (New York: Free Press, 1985), p. 7.

173 *to address them.* Marcy Darnovsky, "'Moral Questions of an Altogether Different Kind': Progressive Politics in the Biotech Age," *Harvard Law & Policy Review* 4:1, p. 102. Winter 2010.

173 *where liberals fear to tread.* Michael Sandel, *Justice: What's the Right Thing to Do?* (New York: Farrar, Straus and Giroux, 2009) p. 243.

177 *called communism.* It is a perverse tribute to the power of the label "science" that so many of the enemies of reason and empirical justification have wanted so badly to bask in its glow, from the radical left (scientific Marxism) to modern quasi-religious cults (Scientology) to creationists (creation science).

180 *DNA might be used.* Caplan A.C. and Moreno J.D., "The Havasu' Baaja Tribe and Informed Consent," *The Lancet*, July 13, 2010. http://www.lancet.com /journals/lancet/article/PIIS0140-6736%2810%2960818-5/fulltext.

181 *any less possible.* I am indebted to Timothy Ferris's discussion of post-modernism in *The Science of Liberty.*

182 *crucial premises of the idea of progress.* Robert Nisbet, *History of the Idea of Progress* (New York: Basic Books, 1980).

182 *generates its strength.* Timothy Ferris, *The Science of Liberty.*

184 *populist an idea as there is.* One eminently practical reason for bioprogressives to commit themselves to the (admittedly often) vague idea of human dignity is that it is enshrined in just about every important international human rights document, among them the Council of Europe's 1997 Convention on Human Rights and Medicine.

184 *"perfectionism."* I am grateful to John Evans for this point.

186 *same body politic.* Information about Tishkoff's remarkable work can be found on her Website, http://www.med.upenn.edu/tishkoff/index.html.

INDEX

ABOUT THE AUTHOR

JONATHAN D. MORENO served on President Obama's transition team and has been a senior staff member for three presidential advisory commissions and on the Bill and Melinda Gates Foundation Bioethics Advisory Board for the Grand Challenges in Global Health Initiative. Author and editor of many seminal books and articles on science and science policy (including *Science Next*, edited with Rick Weiss, and *Mind Wars: Brain Research and National Defense*), he is the David and Lyn Silfen University Professor at the University of Pennsylvania and the editor-in-chief for the Center for American Progress' online magazine, *Science Progress*. He divides his time between Philadelphia and Washington, DC.